Introduction to
Modern Climate Change

ANDREW E. DESSLER

Texas A&M University

CAMBRIDGE
UNIVERSITY PRESS

CAMBRIDGE UNIVERSITY PRESS
Cambridge, New York, Melbourne, Madrid, Cape Town,
Singapore, São Paulo, Delhi, Tokyo, Mexico City

Cambridge University Press
32 Avenue of the Americas, New York, NY 10013-2473, USA

www.cambridge.org
Information on this title: www.cambridge.org/9780521173155

First published 2012

Printed in the United States of America

A catalog record for this publication is available from the British Library.

Library of Congress Cataloging in Publication data
Dessler, Andrew Emory.
Introduction to modern climate change / Andrew Dessler.
p.   cm.
Includes bibliographical references and index.
ISBN 978-1-107-00189-3 (hardback) – ISBN 978-0-521-17315-5 (paperback)
1. Climatic changes.   2. Climatic changes – Government policy.   I. Title.
QC903.D46   2012
551.6–dc23   2011015540

ISBN 978-1-107-00189-3 Hardback
ISBN 978-0-521-17315-5 Paperback

For Alex and Michael

Contents

Preface

Future generations may well view climate change as the defining issue of our time. The worst-case scenarios of climate change are truly terrible, but even middle-of-the-road scenarios portend environmental change without precedent for human society. When future generations look back on our time in charge of the planet, they will either cheer our foresight in dealing with this issue or curse our lack of it.

Yet despite the stakes, the world has done basically nothing to address this risk. The reasons are obvious: The threat of climate change is really a threat to future generations, not the present one, so actions taken by our generation will mostly benefit them and not us. Moreover, such actions may be expensive – reducing emissions means rebuilding our energy infrastructure, and we have no idea how much that will cost. In such a situation, it is easiest to do nothing and wait for disaster to strike – which is why dams are frequently built after the flood, not before. Nevertheless, pushing this problem off onto future generations is a poor strategy. The impacts of climate change are global and mainly irreversible; by the time we have unambiguous evidence that the climate is changing and its impacts are serious, it will be too late to avoid these serious impacts. The only hope that future generations have to avoid serious climate change is us.

I fully believe that the cornerstone of good policy is an electorate that is educated on the issues, and this belief provided me the motivation for writing this book. The goal of the book is to cover the human-induced climate change problem from stem to stern, covering not just the physics of climate change but also the economic, policy, and moral dimensions of the problem. This sets it apart from most other books, which typically do not have a tight focus on human-induced climate change or do not cover the nonscience aspects of the problem.

Such complete coverage of the climate change problem is essential. The science clearly underlies all discussion of the problem, and an understanding of the science is essential to an understanding of why so many people are so worried about it. Climate change, however, is no longer just a scientific problem. Virtually every government in the world now accepts the reality of climate change, and the debate has, to a great extent, moved on to policy questions, including the economic and ethical issues. Thus, one must also understand nonscience aspects of the problem to be truly informed on this issue.

The first seven chapters of the book focus on the science of climate change. Chapter 1 defines the problem and provides definitions of weather, climate, and climate change. It also addresses an issue that most textbooks do not have to address: why the reader should believe this book as opposed to Web sites and other sources that give a completely different view of the climate problem. Chapter 2 explains the evidence that the Earth is warming. The evidence is so overwhelming that there is little argument anymore over this point, and my goal is for the readers come away from the chapter understanding this.

Chapter 3 covers the basic physics of electromagnetic radiation necessary to understand the climate. I use familiar examples in this chapter, such as glowing metal in a blacksmith shop and the incandescent light bulb, to help the reader understand these important concepts. In Chapter 4, a simple energy-balance climate model is derived. It is shown how this simple model successfully explains the Earth's climate as well as the climate of Mercury, Venus, and Mars. The carbon cycle is covered in Chapter 5, and feedbacks, radiative forcing, and climate sensitivity are all discussed in Chapter 6. Finally, Chapter 7 explains why scientists are so confident that humans are to blame for the recent warming that the Earth has experienced.

Chapter 8 begins an inexorable shift from physics to nonscience issues. Chapter 8 discusses emissions scenarios and the social factors that control them, as well as what these scenarios mean for our climate over the next century. Chapter 9 covers the impacts of these changes on humans and on the world in which we live. Chapter 10 covers exponential math. Exponential growth is a key factor in almost all fields of science, as well as in real life. In this chapter, I cover the math of exponential growth and explain the concept of exponential discounting.

Starting with Chapter 11, the discussion is entirely on the policy aspects of the problem. Chapter 11 discusses the three classes of responses to climate change, namely adaptation, mitigation, and geoengineering, and their advantages, disadvantages, and trade-offs. The most contentious arguments over climate change policy are over mitigation, and Chapter 12 discusses in detail the two main policies advanced to reduce emissions: carbon taxes and cap-and-trade systems.

Chapter 13 provides a short history of climate science and a history of the political debate over this issue, including discussions of the United Nation's Framework Convention on Climate Change and the Kyoto Protocol. Finally, Chapter 14 pulls the last three chapters together by discussing methods of deciding which of our options we should adopt, particularly given the pervasive uncertainty in the problem.

Overall, it should be possible to cover about one chapter in three hours of lecture. This makes it feasible to cover the entire book in one 15-week semester. At Texas A&M, the material in this book is being used in a one-semester class for nonscience majors that satisfies the University's science distribution requirement. Thus, it is appropriate for undergraduates with any academic background and at any point in their college career.

Any serious understanding of climate change must be quantitative. Therefore, the book assumes a knowledge of simple algebra. No higher math is required. The book also assumes no prior knowledge of any field of science, just an open mind and willingness to learn. To aid in the student's development of a numerate understanding of the climate, there are quantitative questions at the end of many of the chapters, and every chapter also has more open-ended, qualitative questions. In addition, there is a chapter summary at the end of each chapter that reviews and summarizes the most important take-away messages from the chapter. A list of important terms is also provided at the end of each chapter.

This is not a book of advocacy. This is not to say that I do not have opinions. I do, and strong ones. I recognize, though, that shrill advocacy is frequently less effective than a dispassionate presentation of the facts. Thus, my strategy in this book is to just explain the science and then lay out the possible solutions and trade-offs among them. I firmly believe that an unbiased assessment of the facts will bring the majority of people to see things the way I do: that climate change poses a serious risk and that we should therefore be heading off that risk by reducing our emissions of greenhouse gases.

Every year that our society does nothing to address climate change makes solving the problem both harder and more expensive. I am still optimistic, though, because problems often appear intractable at first. In the 1980s, as evidence mounted that industrial chemicals were depleting the ozone, it was not at all clear that we could avoid serious ozone depletion at a reasonable cost. The chemicals causing the ozone loss, namely chlorofluorocarbons, played an important role in our everyday life – in refrigeration, air conditioning, and many industrial processes – just like the main cause of climate change, fossil fuels, also play an important role in our society. But the cleverness of humans prevailed. A substitute chemical was developed and it seamlessly and cheaply replaced the ozone-destroying halocarbons – all at a cost so low that hardly anyone noticed when the substitution took place.

I realize that solving the climate change problem will be much harder than solving the ozone depletion problem – how much harder, no one knows. I am confident, though, that the ingenuity and creativeness of humans is so great that we can solve this problem without damaging our standard of living. However, there is only one way to find out – and that is to try to do it.

Acknowledgments

Many people have helped me write this book. I thank Rob Korty, John Nielsen-Gammon, Jerry North, R. Saravanan, Russ Schumacher, Debbie Thomas, Andrew Wang, and Shari Yvon-Lewis, for reading and commenting on various parts of the book. Much of this book was written while I was on faculty development leave from Texas A&M University during the fall of 2010, and I thank the university for this support.

1 An introduction to the climate problem

This chapter begins our trip through the climate problem by defining what climate and climate change are, and how we use latitude and longitude to describe locations on the Earth. This chapter also addresses a question that most textbooks do not have to address: Why you should believe it.

1.1 What is climate?

The American Meteorological Society defines *climate* as

> The slowly varying aspects of the atmosphere–hydrosphere–land surface system. It is typically characterized in terms of suitable averages of the climate system over periods of a month or more, taking into consideration the variability in time of these averaged quantities.

Mark Twain, in contrast, famously summed it up a bit more concisely:

> Climate is what you expect; weather is what you get.

Put another way, *weather* refers to the actual state of the atmosphere at a particular time. We are referring to the weather when we say that the low and high temperatures on August 8, 1993 in College Station, TX, were 24 °C and 37 °C, respectively, and there was no precipitation.

Climate, in contrast, is used for a statistical description of the weather over a period of time, usually a few decades. It could include the average temperature, for example, as well as a measure of how much the temperature varies about this average value, such as the record high and low temperatures. Figure 1.1 shows the distribution of daily high and low temperatures in August in Fairbanks, AK between 1975 and 2008. It shows, for example, that the most likely high temperature is 23 °C, which occurred on approximately 5% of the days during this period. It also shows that extremes occur less frequently; for example, the probability of high temperatures above 30 °C or below 8 °C are quite small. The climate tells us only the range of probable conditions on a particular day; it contains no information about what the temperature was on August 8, 1993.

In this book I frequently use the Celsius scale, the most widely used temperature scale in the world (the Fahrenheit scale more familiar to U.S. readers is only used in the United States and a few other countries). Celsius is also used by scientists, and because this book is foremost a science textbook, I have adopted the Celsius scale. In Chapter 3 I discuss the Kelvin scale, which is also widely used by scientists.

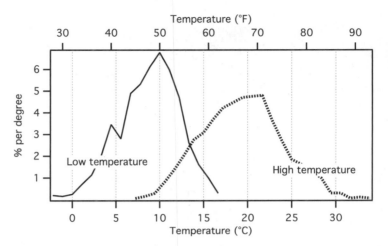

Fig. 1.1 Frequency of occurrence of daily high and low temperatures in August in Fairbanks, AK between 1975 and 2009 (data obtained from the National Climatic Data Center; http://cdo.ncdc.noaa.gov).

It is useful to remember that the freezing and boiling temperatures for water in the Celsius scale are 0 °C and 100 °C, respectively. In the Fahrenheit scale, these temperatures are 32 °F and 212 °F. Room temperature is approximately 72 °F or 22 °C. And to convert from Fahrenheit to Celsius, use the equation C = (F − 32) × 5/9; to convert from Celsius to Fahrenheit, use the equation F = C × 9/5 + 32.

Why do we care about weather and climate? Weather, on one hand, is important for making short-term decisions. For example, should you take an umbrella when you leave the house tomorrow? To answer this question, you don't care at all about the average precipitation for the month, but rather whether it is going to rain *tomorrow*. If you are going skiing this weekend, you care about whether new snow will fall before you arrive at the ski lodge and what the weather will be while you are there. You do not care how much snow the lodge gets on average.

Climate, on the other hand, is more important for long-term decisions. If you are looking to build a vacation home, you are interested in finding a place that frequently has pleasant weather – you are not particularly interested in the weather on any specific day. If you are building a ski resort, you want to place it in a location that, on average, gets enough snow to produce acceptable ski conditions. You do not care if snow is going to fall on a particular weekend, or even what the total snowfall will be for particular year; you are interested in the long-term statistics of snowfall.

A good example of the importance of both the climate and weather can be found in the planning for D-Day, the invasion of the European mainland by the Allies during World War II. The invasion required thousands of Allied troops to be transported onto the beaches of Normandy, along with enough equipment that they could establish and hold a beachhead. As part of this plan, thousands of Allied paratroopers were to be dropped into the French countryside in the middle of the night before the beach landing in order to capture strategic towns and bridges near the landing zone, thus hindering a German counterattack.

There were important weather requirements for the invasion. The nighttime paratrooper drop required a cloudless night as well as a full moon so that the paratroopers would be able to land safely and on target, and then achieve their objectives – all before dawn. The sky had to remain clear during the next day so that air support could see targets on the ground. For tanks and other heavy equipment to be brought onshore required firm, dry ground, so there should be no heavy rains just prior to the invasion. Furthermore, the winds could not be too strong, because high winds generate big waves that create problems for the Navy, particularly for the small landing craft that would ferry infantry to the beaches.

Given these and other weather requirements, Allied analysts studied the climate of the candidate landing zones to find those beaches where the required weather conditions occurred most frequently. The beaches of Normandy were ultimately selected because of its favorable climate and other tactical considerations.

Once the landing location had been selected, the exact date of the invasion would have to be selected. For this, it would not be the climate that mattered but rather the weather on a particular day. Operational factors such as the phase of the tide and the moon provided a window of 3 days for a possible invasion: June 5, 6, and 7, 1944. June 5 was initially chosen, but on June 4, as ships began to head out to sea, bad weather set in at Normandy and General Dwight D. Eisenhower made the decision to delay the invasion. On the morning of June 5, chief meteorologist J. M. Stagg forecasted a break in the weather – and Eisenhower decided to proceed. Within hours an armada of ships set sail for Normandy. That night, hundreds of aircraft carrying thousands of paratroopers roared overhead to the Normandy landing zones.

The invasion began just after midnight on June 6, 1944 when British paratroopers seized a bridge over the Caen Canal. At dawn, 3,500 landing craft carrying tens of thousands of soldiers hit the beaches. Stagg's forecast was accurate and the weather was good, and despite ferocious casualties, the invasion succeeded in placing an Allied army on the European mainland. This was a pivotal battle of World War II, marking a key turning point in the war. And analyses of both weather and climate played a key role in the success of this mission.

Temperature is the parameter most often associated with climate, and it is something that directly affects the well-being of the Earth's inhabitants. The statistic that most frequently gets discussed is average temperature, but temperature extremes also matter. For example, it is heat waves – prolonged periods of excessively hot weather – rather than normal high temperatures that kill people. In fact, heat-related mortality is the leading cause of weather-related death in the United States (it kills many more people than cold-related mortality). And the numbers can be staggering: In August of 2003, a severe heat wave in Europe lasting several weeks killed tens of thousands of people.

Precipitation rivals or even exceeds temperature in its importance to humans, because human life without fresh water is impossible. As a result, precipitation is almost always included in any definition of climate. Total annual precipitation is obviously an important part of the climate of a region. However, the distribution of this rainfall throughout the year also matters. Imagine, for example, two regions that get the same total amount of rainfall each year. One region gets the rain evenly distributed throughout the year, whereas the other region gets all of the rain in 1 month, followed by 11 rain-free months. The environment of these two regions would be completely different. Where the rain falls continuously throughout the year, we would expect a green, lush environment. Where there are long rain-free periods, in contrast, we expect something that looks more like a desert.

Other aspects of precipitation, such as its form (rain vs. snow), are also important. In the U.S. Pacific Northwest, for example, snow that accumulates in the mountains during the winter melts during the following summer, thereby providing fresh water to the environment during the otherwise dry summers. If warming causes wintertime precipitation to fall as rain rather than snow, then it will run off immediately and not be available during the following summer. This can lead to water shortages during the summer.

As these examples show, climate includes many environmental parameters. What part of the climate matters will vary from person to person, depending on how he or she relies on the climate. The farmer, ski resort owner, resident of Seattle, and Dwight D. Eisenhower are all interested in different meteorological variables, and thus may care about different aspects of the climate. But make no mistake: We all rely on the stability of our climate. In particular, food production and freshwater availability, two of the most important things we rely on to survive, are greatly affected by the climate. I will discuss this in greater depth when I explore climate impacts in Chapter 9.

A final difference between weather and climate is how easy they are to determine. Measuring the weather is pretty easy – just walk outside and look around.[1] If you need a higher level of accuracy, you can buy reasonably cheap instruments to measure the temperature, precipitation, or any other variable of interest. Climate, in contrast, is much harder to measure; it requires the gathering of decades of data so that we have sufficiently good, robust statistics, such as I plotted in Figure 1.1. I will discuss this challenge in greater detail in Chapter 2.

1.2 What is climate change?

The climate change that is most familiar is the seasonal cycle: the progression of seasons from summer to fall to winter to spring and back to summer, during which most locations experience significant temperature variations. Precipitation can also vary by season. In fact, almost any climate variable can vary over the course of the year.

The concern in the climate change debate – and in this book – is with long-term climate change. The American Meteorological Society defines the term *climate change* as follows:

> [It is] any systematic change in the long-term statistics of climate elements (such as temperature, pressure, or winds) sustained over several decades or longer.

---

[1] There are, of course, siting issues in measuring the weather. Depending on your location, the weather you measure when you walk outside may not be terribly representative of the weather of the larger areas.

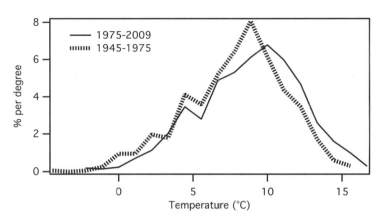

Fig. 1.2 Frequency of occurrence of daily low temperature in August in Fairbanks, AK for two periods, 1945–1975 and 1975–2009 (data obtained from the National Climatic Data Center; http://cdo.ncdc.noaa.gov).

In other words, we can compare the climate for one period against the climate for another period, and if the statistics have changed, then we can say that the climate has changed.

Thus, we are interested in whether today's climate (defined over the past few decades) is different from the climate of a century ago, and we are worried that the climate at the end of the 21st century will be quite different from that of today. As an example, Figure 1.2 plots the August minimum temperature in Fairbanks, AK for two periods, 1945–1975 and 1975–2009. The distribution of daily minimum temperature has clearly shifted, from an average of 7.6 °C in the early period to an average of 8.5 °C in the later period. In addition to the shift in average temperatures, we can see that warm temperatures became more frequent and cold temperatures became less frequent. It should also be noted that there is no information on the cause of the change in this plot – it may be due to global warming or one of any number of other physical processes. All we have identified here is a shift in the climate.

The increase in daily minimum temperature is only 0.9 °C, and it might be tempting to dismiss this as unimportant. However, as I discuss in Chapter 9, seemingly small changes in climate are associated with significant impacts on the environment. Do not dismiss such a change lightly.

In Chapter 2, we will pick up this theme and look at data to determine if the climate is indeed changing. Before we get to that, though, there are two things I need to cover. First is the coordinate system I will be using in this book. The second is a more general discussion about why you should believe the science in this textbook.

1.3 A coordinate system for the Earth

I will be talking a lot in this book about the Earth, so it makes sense to describe the terminology used to identify particular locations and regions on the Earth.

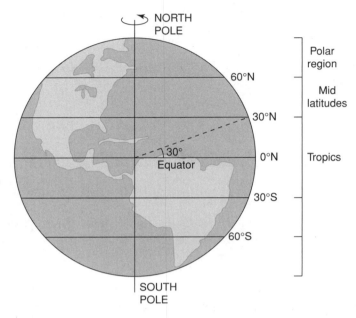

A schematic showing of latitude.

To begin, the *equator* is the line on the Earth's surface that is halfway between the North and South Pole, and it divides the Earth into a northern hemisphere and a southern hemisphere. The *latitude* of a particular location is the distance in the north–south direction between the location and the equator (Figure 1.3), measured in degrees. Latitudes for points in the northern hemisphere have the letter *N* appended to them, with *S* appended to points in the southern hemisphere. Thus, 30 °N means a point on the Earth that is 30 ° north of the equator, whereas 30 °S means the same distance south of the equator.

The *tropics* are conventionally defined as the region from 30 °N to 30 °S, and this region covers half the surface area of the planet. The *mid-latitudes* are usually defined as the region from 30 ° to 60 ° in both hemispheres, and these regions occupy roughly one third of the surface area of the planet. The *polar regions* are typically defined to be 60 ° to the pole, and these regions occupy the remaining one sixth of the surface area of the planet. The North and South Poles are located at 90 °N and 90 °S, respectively.

Latitude gives the north–south location of an object, but to uniquely identify a spot on the Earth you also need to know the east–west location. That is where *longitude* comes in (Figure 1.4). Longitude is the angle in the east or west direction, from the *prime meridian*, a line that runs from the North Pole to the South Pole through Greenwich, England, and is arbitrarily defined to be 0 ° longitude. Locations to the east of the prime meridian are in the eastern hemisphere and have the angle appended with the letter *E*, whereas locations to the west are in the western hemisphere and have the letter *W* appended. In both directions, longitude increases to 180 °, where east meets west.

Fig. 1.4   A schematic showing of longitude.

Together, latitude and longitude identify the location of every point on the planet Earth. For example, the Department of Atmospheric Sciences of Texas A&M University is located (approximately) at 30.6 °N, 96.3 °W. Knowing your location can literally be a matter of life and death – shipwrecks, wars, and other miscellaneous forms of death and disaster have occurred because people did not know where they were. Luckily for us, for around $100 you can buy a GPS (global positioning system) receiver that will give your latitude and longitude to within a few feet.

1.4 Why you should believe this textbook

I now have to address an issue that generally does not come up in a college textbook: why you should believe it. Students in most classes accept without question that the textbook is correct. After all, the author is probably an authority on the subject, the publisher has almost certainly reviewed the material for accuracy, and the instructor of the class, someone with knowledge of the field, selected that textbook. Given those facts, it seems reasonable to simply assume that the information in the textbook is basically correct.

But climate change is not like every other subject. If you do a quick Internet search, you can likely find a Web page that disputes almost any claim made in this textbook. Your friends and family may not believe that climate change is a serious problem, or they may even believe it is a hoax. You may agree with them. This book will challenge many skeptical viewpoints, and you may face the dilemma of whom to believe.

This situation brings up an important and interesting question: How do you determine whether or not to believe a scientific claim? If you happen to know a lot about an issue, you can reach your own conclusions on the issue. However, no one can be an expert on every subject; for the majority of issues, on which you are not an expert, you need a shortcut.

One type of shortcut is to rely on your firsthand experience about how the world works. Claims that fit with your own experience are easier to accept than those that run counter to it. People do this sort of evaluation all the time, usually unconsciously. Consider, for example, a claim that the Earth's climate is stable. In your lifetime, climate has changed very little, so this seems like a plausible claim. However, a geologist who knows that dramatic climate shifts are responsible for the wide variety of rock and fossil deposits found on Earth might regard the idea of a stable climate as ludicrous, but in turn might be less likely to accept a human origin for climate change. The problem with relying on firsthand experience about the climate is that our present situation is unique – people have never changed the composition of the global atmosphere as much or as fast as is currently occurring. Thus, whatever the response will be, it may be outside the realm of our and the Earth's experiences.

Another type of shortcut is to rely on your values: You can accept the claims that fit with your overall world view while rejecting the claims that do not. For example, consider the scientific claim that second-hand smoke has negative health consequences. If you are a believer in unfettered freedom, you might choose to simply reject this claim out of hand because it implies that governments should regulate smoking in public places to protect public health.

Yet another shortcut is to rely on an *opinion leader*. Opinion leaders are people that you trust, because they appear to be authoritative or because you agree with them on other issues. They might include a family member or influential friend, a media figure such as conservative talk show host Rush Limbaugh or comedian Jon Stewart, or an influential politician such as Barack Obama or George W. Bush. In the absence of a strong opinion of your own, you can simply adopt the view of your opinion leaders. The problem with this approach is that there is no guarantee that the opinion leaders have a firm grasp of the science.

The best approach is to rely on the opinion of experts. When the relevant experts on some subject have high confidence that a scientific claim is true, that is the best indication we have that the claim actually is true. This is not just my view; I am willing to bet it is something you believe in, too. If a friend tells you that she thinks she may be sick, what would you recommend? Your recommendation is likely to be that she should go see a doctor – and not just any doctor, but one who is an expert in that particular ailment.

This is also the view of the U.S. legal system. Many court cases involve questions of science (e.g., what was the cause of death, does a particular chemical cause cancer, does a DNA sample match the defendant). To settle those cases, the court will frequently turn to expert witnesses. These expert witnesses are, as their name suggests, experts on the matter that they are testifying about, and they provide relevant expertise to the court to help evaluate the important scientific questions that a case may revolve around.

To be an expert witness, one must demonstrate expertise in a particular subject. I have served as an expert witness on climate change in lawsuits over the permitting of coal-fired power plants, and the court qualifies me as an "expert" by using my research in climate change as well as the textbooks I have authored as evidence. Other members of my department have served as expert witnesses in lawsuits that have weather-related aspects to them. For example, in a lawsuit involving a car wreck, an expert on weather may testify about the weather conditions at the time, about the visibility, about the possibility that there was ice on the ground, and so on.

It should be emphasized that one must demonstrate specific, recent expertise in the exact area under consideration to be an expert witness. Showing expertise in general technical matters or in a related field is not sufficient. For example, one might consider anyone with a Ph.D. in physics to have a credible opinion about the science of climate change. This is not so, and a person with a Ph.D. in physics without specialized knowledge of the climate would not be qualified to be an expert on matters of climate. That also goes for weather forecasters – climate and weather are different, and being an expert in weather would not qualify someone to be an expert witness on climate. The reverse is also true, so I, despite being a professor of atmospheric sciences, would not qualify as an expert in weather. The requirement for the expertise to be recent rules out those who were experts, say, a decade ago but who have not kept up with the latest discoveries in the field.

There are many more examples that demonstrate that, as individuals and as a society, we have decided that expertise counts when one is evaluating competing claims on matters of science. That is probably a good thing, too, because on a planet with almost 7 billion people, you can always find someone who will contest any claim, no matter how well established it is. For example, it would be relatively easy to find someone somewhere who would dispute the claim that cigarettes cause health problems. So if everyone's opinion counted equally, then it would be impossible to ever settle any dispute over a scientific claim – even one as simple as whether the Earth goes around the Sun.

Nonetheless, you also know that experts are not all equal. If one of your friends needs to see an endocrinologist for treatment of a serious endocrine disorder, you are not going to recommend that he open the yellow pages and call the first one he finds. Rather, you will suggest that he try to find the best one, perhaps by asking friends, family, or their family doctor for recommendations, or do research online to find someone with outstanding credentials.

For important medical decisions, though, even finding a doctor you trust is not enough. After all, anyone – even the most trusted expert – can make a mistake. Moreover, some people have biases that may be undetectable. One way to gain additional confidence in a particular diagnosis is to get a second opinion. If you have the time and resources, you may even get more opinions. If all of the experts agree, then you would have justifiably high confidence that the recommendations are the best advice that modern medicine can provide.

Climate change is really no different. It is obvious that the relevant experts are the community of climate scientists. However, there are thousands of climate scientists out there, so which ones should we to listen to? One approach would be to ask all of the world's climate scientists what they think – and if the vast majority agree on a particular point, then we can have high confidence that point of view is correct.

This is, in fact, what has already been done. In 1988, as nations began to acknowledge the seriousness of the climate problem, the Intergovernmental Panel on Climate Change (IPCC) was formed. The IPCC assembles large writing teams of scientific experts and has them write, as a group, a report detailing what they know about climate change and how confidently they know it. The reliance on large writing groups reduces the possibility that the erroneous opinions of an individual or a small group make it into the report, much like getting multiple opinions in medicine reduces the chance of a bad recommendation.

To further minimize the possibility that the group of scientists writing the report are biased in some direction, the scientists making up the writing teams are not drafted by a single person or organization; they are nominated by the world's governments. Thus, the only way the IPCC's writing groups would be biased in some direction is if all of the world's governments nominated biased individuals. This seems very unlikely, particularly because some of the world's governments are very concerned about climate change whereas others would be very happy if climate change disappeared completely as a political issue.

After being written by experts, the IPCC's reports are then reviewed by other expert scientists, and they undergo a public review and a separate review by the world's governments. In the end, the IPCC's reports[2] are widely regarded as the most authoritative statements of scientific knowledge about climate change, and as such they carry enormous weight in both the scientific and policy communities. The reports are not perfect (no complex document written by humans can be), but they are really quite good. In 2007, the IPCC shared the Nobel Peace Prize in recognition of its work on the climate.

In addition to the IPCC's reports, you can also examine reports from other assessment organizations, such as the United States National Academy of Sciences. Or you can look at the statements put out by the scientific societies that climate experts belong to. For example, in October of 2009, a collection of U.S. scientific organizations sent a letter to the U.S. Senate stating that climate change is a serious problem facing the entire human race and that emissions of greenhouse gases have to be dramatically reduced for us to avoid the most severe impacts.[3] Signatories of this letter include the American Association for the Advancement of Science, the American Chemical Society, the American Geophysical Union, the American Institute of Biological Sciences, the American Meteorological Society, the American Society of Agronomy, the American Society of Plant Biologists, the American Statistical Association, the Association of Ecosystem Research Centers, the Botanical Society of America, the Crop Science Society of America, the Ecological Society of America, the Natural

---

[2] These reports can be downloaded (available at http://www.ipcc.ch).
[3] This letter is available online (see http://www.aaas.org/news/releases/2009/media/1021climate_letter
   .pdf).

Science Collections Alliance, the Organization of Biological Field Stations, the Society for Industrial and Applied Mathematics, the Society of Systematic Biologists, the Soil Science Society of America, and the University Corporation for Atmospheric Research. Comparable non-U.S. scientific organizations in other countries have also endorsed the mainstream view of the science of climate change.

To understand the public debate over climate, consider the following scenario. Imagine that you are a captain of industry – you are the owner of a big industrial complex that makes a great deal of stuff that is sold to make piles of money. Then some liberal egghead scientists come along and tell you that your industry is harming the environment. You have two choices: You can meekly accept regulations on your industry, which will cut into profits and destroy the price of your stock, or you can fight back by attacking the science.

You will, of course, fight back – just like every other captain of industry has done when science points out that something she is doing is harming the environment. Thus, it should come as no surprise that the IPCC has come under withering attacks by those opposed to political action to address climate change. This is out of necessity: Those opposed to action on climate change must do this or they cede any scientific basis for opposing action.

One argument frequently made is that there is, in fact, wide disagreement on the science of climate change among climate scientists. As evidence, they will point to Internet petitions and various lists of scientists that dispute the mainstream view. However, a close evaluation of the dissident scientists on these lists and petitions reveals that in almost all cases they should not be considered experts on climate. Although many of the individuals on the lists have technical degrees, and some even have doctorates, their specific training does not include climate change. They would never qualify as an expert witness in a lawsuit on climate change; they do not have the background to teach a college-level course on the material; and we would never trust the diagnosis provided by a doctor with an equivalent expertise.

The only reason that advocates put such transparently unqualified people forward as experts is that legitimate experts with the desired opinions are not available. Thus, the lack of credentials of those on the petitions and lists actually underscores the strong agreement among the relevant experts on the science of climate change.

A second claim we may hear is that climate scientists are manufacturing a crisis to benefit themselves. If climate is a crisis, so the argument goes, then more research funding will flow into the field, the prestige of climate scientists will increase, and scientists will be able to implement their preferred social policies.

There is, in fact, no evidence to support this argument. Rather, this argument relies on the listener's simply accepting the obviousness of the claim that an entire scientific field has been corrupted by the enticement of research funding. What is often lost in this discussion is that most, if not all, scientific fields have this same incentive. Biomedical fields could invent a new disease or a cure for an existing one to increase funding; physicists studying solid-state physics could invent a discovery that could lead to much faster computers; and space physicists could invent evidence of life elsewhere in the solar system. All of these "discoveries" would increase funding and interest in the particular field of interest.

Nonetheless, for many reasons, such widespread fraud by an entire scientific community has never occurred in science. First and foremost, there is a coordination problem: How do you get everyone to go along with the fraud? The answer is that you can't. A scientific field such as climate science is a large, diverse, and intensely competitive endeavor, and the desire to outthink one's peers and show that one is smarter than they are is much greater than the incentive to cooperate in this type of fraud. The reason for this is that success in science is achieved by impressing one's colleagues. One of the best ways to do this is to overturn conventional wisdom, either by showing that previous scientific results are wrong or by suggesting a new theory that fits the data better. Because this is so beneficial to the reputation of the individual scientist, it provides a strong incentive against participating in any conspiracy.

Second, the incentives in science do not support such a conspiracy. Money from grants does not generally go into the pockets of the researchers. Most scientists are employees of federal or state governments or private universities, and the amount of money they can pay themselves off grants is tightly regulated. Rather, money from grants generally goes to buy equipment or pay for graduate students. Thus, research funding provides a very weak incentive to cheat.

Finally, the entire underlying premise of the "climate science is corrupt" narrative is questionable. The premise is that, by suggesting that human effects on the climate are well understood, the field gets more research funding than it would otherwise. However, history suggests that whenever a field reaches a conclusion on a problem, funding for further research on that problem goes down. For example, after the ozone depletion problem was solved in the mid-1990s, the funding for subsequent research rapidly dropped. Today, there are few scientists working on the problem. By saying that they understand the climate system pretty well, members of the climate science community are *not* helping their funding. They would do better if they claimed that there was no consensus on why the climate is warming. In that case, it would almost certainly be a high priority for most policy makers to fund research into determining the cause of the warming.

However, although there has never been widespread fraud by an entire scientific community, there are cases in which advocates opposed to political action have falsely tried to cast doubt on the science. For example, leaked documents from tobacco companies have shown that these companies engaged in a well-designed campaign to generate scientific doubt in the public's mind about the health effects of cigarette smoking. Those opposed to action on climate change have adopted many of the same tactics.[4]

Because of this, we should be leery of the argument that "the experts can't be trusted." It goes against both common sense and our experience in the real world, and it should only be accepted if extraordinary evidence is provided. In the climate change debate, such evidence is clearly lacking.

The science in this book follows the IPCC's reports. That, in a nutshell, is why you should believe this book. The alternative views on climate change you might see or hear, such as those from friends or the Internet, do not come from a process as credible as the IPCC's, and therefore do not have the same standing. It should be

---

[4] See the description given by the Union of Concerned Scientists (2007).

emphasized that this does not mean the IPCC's reports are correct – any scientific claim is at risk of being overturned by future research. Nevertheless, the IPCC does accurately represent what the relevant experts think about the science, which is the best guide there is for nonexperts.

1.5 Chapter summary

- *Weather* refers to the exact state of the atmosphere at a point in time; *climate* refers to the statistics of the atmosphere over a period of time, usually several decades in length or longer.
- Climate change refers to a change in the statistics of the atmosphere over decades. Such statistics include not just the averages but also the measures of the extremes – how much the atmosphere can depart from the average.
- Temperatures expressed in this book are in degrees Celsius; conversion from Fahrenheit can be done with this equation: $C = (F - 32) \times 5/9$.
- Any position on the surface of the Earth can be described by a latitude and longitude; the tropics cover the region from $30\,°N$ to $30\,°S$; mid-latitudes cover the region from $30\,°$ to $60\,°$ latitude; and the polar regions cover from $60\,°$ to $90\,°$ latitude.
- In our society, we frequently rely on experts for advice on highly specialized or technical fields. For climate change, the IPCC reports represent the opinion of the world's experts, and the science described in this book reflects the IPCC's scientific views.

Additional reading

S. R. Weart, *The Discovery of Global Warming*, 2nd ed. (Cambridge, MA: Harvard University Press, 2008). This is a highly readable and accessible history of major developments in the science of climate change, from the 19th century through the formation of the modern consensus about the reality and predominant human cause of recent climate change as expressed in the 2001 IPCC report (available online at http://www.aip.org/history/climate/index.htm).

S. R. Weart, "Changing the Climate . . . of Public Opinion," *APS News* 15(2) (2006): 12. This article gives a concise discussion of scientific assessments and the IPCC.

J. Houghton, "Madrid 1995: Diagnosing Climate Change," *Nature* 455 (2008): 737–738. This is a short discussion of how IPCC conclusions are arrived at. You will see here why they are so reliable (available online at http://www.nature.com/nature/journal/v455/n7214/full/455737a.html).

N. Oreskes and E. M. Conway, *Merchants of Doubt: How a Handful of Scientists Obscured the Truth on Issues from Tobacco Smoke to Global Warming* (London: Bloomsbury Press, 2010). This is an important book about how deception is used to mislead the public on matters ranging from the risks of smoking to ozone depletion to the reality of global warming.

Union of Concerned Scientists, *Smoke, Mirrors, and Hot Air: How ExxonMobil Uses Big Tobacco's Tactics to Manufacture Uncertainty on Climate Science* (Cambridge, MA: UCS, January 2007). This is a description of how the tactics employed by the tobacco companies to cast doubt on the science of the health impacts of smoking are now being used by oil companies to cast doubt on the science of climate change (available online at http://www.ucsusa.org/assets/documents/global_warming/exxon_report.pdf).

Terms

Climate
Climate change
Equator
Latitude
Longitude
Mid-latitudes
Opinion leader
Polar region
Prime meridian
Tropics
Weather

Problems

1. Determine the latitude and longitude of the White House, the Kremlin, the Pyramids of Giza, and the point on the opposite side of the Earth to where you were born. Use an online tool (e.g., Google Earth) or an atlas (which you can find in any library).

2. a) Convert the following temperatures from degrees Fahrenheit to degrees Celsius:

   300, 212, 70, 50, 32, and 0 °F

   b) Convert the following temperatures from degrees Celsius to degrees Fahrenheit:

   150, 100, 70, 50, 0, and −10 °C

3. a) The temperature increases by 1 °C. How much does it increase in degrees Fahrenheit?

   b) The temperature increases by 1 °F. How much does it increase in degrees Celsius?

4. What temperature has a numerical value that is the same in degrees Celsius as it is in degrees Fahrenheit?

5. Find a two-digit temperature in degrees Fahrenheit for which, if you reverse the digits, you get that same temperature in degrees Celsius (e.g., find a temperature, such as 32 °F, for which the Celsius equivalent would be 23 °C; this example, of course, does not work).

6. Why do you believe that smoking causes cancer? (If you do not believe this, then why do you believe that smoking does not cause cancer?) What would be required to get you to adopt the opposing view?

7. Find two friends who have strong but opposing views of climate change.
   a) Ask both of them why they believe what they do and what would be required for them to adopt the opposing view. It is important to understand where their views come from; if they argue, say, that glaciers are retreating or not, find out where they get their facts.
   b) Which of these positions appears more credible? Why?
   c) Can you use their views on climate change to predict their views on other issues (abortion, gun control) and their political affiliation?

8. Practice reading a graph. These questions all refer to Figure 1.1.
   a) What is the fraction of days that have a high temperature of 15 °C?
   b) What is the fraction of days that have a low temperature of 10 °C?
   c) Below what temperature is the probability for the daily high temperature zero?
   d) Above what temperature is the probability for the daily high temperature zero?
   e) For what temperature is there an equal probability of having a daily high and low?

2

Is the climate changing?

In Chapter 1, I defined climate change as a change in the statistics of the weather (changes in averages and extremes of temperature and precipitation and other meteorological parameters of interest) over a period of several decades. In this chapter, I address the question of whether the Earth's climate is currently changing and how it has changed in the past. We will see the overwhelming evidence that the climate is indeed changing – and that it has changed significantly over the Earth's entire history. In Chapter 7, I will discuss what causes climate change and whether humans are to blame for the present warming.

Although there are many statistics that we could examine, the discussion in this chapter focuses on temperature for two reasons. First, as I will discuss in Chapter 4, the most direct impact from the addition of greenhouse gases to the atmosphere is a warming planet. Changes in other variables, such as precipitation or sea level, arise as a response to the temperature change. Second, we have the best data for temperature. The technology for measuring it is centuries old, and people have been measuring and recording the temperature with reasonable global coverage since the middle of the 19th century. In addition to direct measurements of temperature, there are other techniques, such as studying the chemical composition of rocks, which allow us to indirectly infer the temperature of the Earth over nearly its entire 4.5-billion-year history.

Rather than analyze temperature directly, however, we will instead analyze *temperature anomalies*. A temperature anomaly is the difference between the actual temperature and a reference temperature, usually an average over a previous multi-decadal period. Figure 2.1 shows the temperature anomaly for December 2009. In this case, the anomaly at each location is calculated by taking the December 2009 temperature and subtracting from it the average temperature at that location for all Decembers between 2000 and 2009. Thus, positive anomalies mean that the temperature in December 2009 was warmer than the December 2000–2009 average at that location, whereas negative anomalies mean the temperatures are cooler.

An example of an anomaly calculation

Imagine that you want to catalog the height of a group of your friends. One approach is to simply record everyone's actual height: 5 ft, 3 in.; 6 ft, 1 in.; 5 ft, 6 in.; 5 ft, 9 in.; and so on. Alternatively, you could record their height anomalies. To convert their heights to anomalies, you subtract a reference height, say, 5 ft, 6 in., from their actual height. So someone who is 5 ft, 3 in. would have a height anomaly of −3 in., whereas someone who is 5 ft, 9 in. would have a height anomaly of +3 in. Thus, the height anomalies (with respect to the reference height of 5 ft, 6 in.) for the group are −3, +7, 0, +3, and so on.

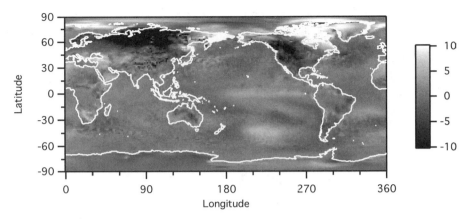

Fig. 2.1 The monthly surface temperature anomaly in December 2009 in degrees Celsius. The reference temperature for the anomaly calculation is calculated at each location as the average of the December temperatures from 2000 to 2009 at that location (data obtained from the ECMWF-interim reanalysis). (See Color Plate 2.1.)

Why use anomalies rather than absolute temperature in these figures? The main reason is that absolute temperature can vary sharply over short distances, such as between a city and a nearby rural area, or between two nearby sites at different altitudes. You may have noticed this, for example, if your car displays outside temperature on its dashboard. As you drive a few miles, you will frequently see the temperature vary by a few degrees, particularly if you are driving into and out of a city. Anomalies, however, are constant over much longer distances: If it is a degree warmer than average in a city, then it is probably a degree warmer than average a few kilometers away from the city, even if the absolute temperatures are different by a few degrees. This makes the calculation of anomalies more accurate by requiring a less dense measurement network. Thus, just about all of the data shown in this chapter are in the form of anomalies.

Finally, we will focus on the global annual average temperature. The reason is that local climate variations are frequently balanced by opposite variations elsewhere. So if one region is undergoing a heat wave, there is likely another region that is undergoing a cold wave. For example, Figure 2.1 shows the distribution in latitude and longitude of the temperature anomaly for December 2009. The cold temperatures over North America and Northern Europe are balanced by warm temperatures that cover much of the rest of the globe. You would be quite misled if you concluded that the Earth was cold because of cold temperatures found over Europe.

2.1 Recent climate change

2.1.1 Surface thermometer record

People have been measuring the local air temperature at locations all over the globe for centuries. By combining these measurements, scientists have constructed an estimate

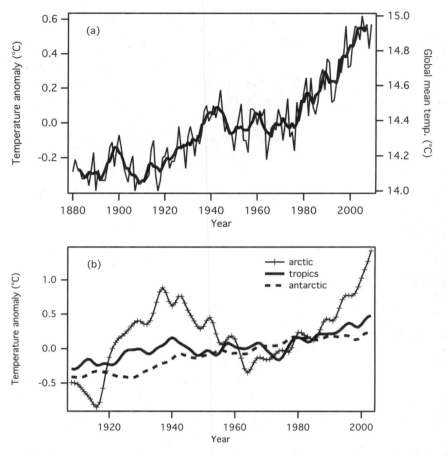

Fig. 2.2 Global annual average temperature anomalies: (a) The right-hand axis shows approximate global average temperature. The gray line is a smoothed time series. (b) Smoothed temperature time series for three different regions of the planet: the Arctic (64 °N–90 °N), the tropics (24 °N–24 °S), and the Antarctic (64 °S–90 °S). In both plots, the reference temperature used in calculating the anomalies is the 1951–1980 average (data are from the NASA GISS Surface Temperature Analysis, or GISTEMP, product – see http://data.giss.nasa.gov/gistemp/).

of the average surface temperature of the Earth over the past 150 years, and the time series of global annual temperature anomalies is plotted in Figure 2.2a.

As should be obvious, the data show the Earth is warming, and the past decade (2000–2009) is the warmest on the record. From 1906 to 2005, the average surface temperature of the Earth rose by 0.74 °C. Most of this increase occurred in two distinct periods, from 1910 to 1945 and from 1976 to the present, with a period of little change in between. There are also many shorter-term bumps and wiggles over the century. The rate of warming over the past 50 years is 1.3 °C per century, which is approximately twice the rate for the entire century.

The warming of the planet is not uniform, though, and some regions have experienced more warming than others. In general, land areas have warmed more than the oceans. Figure 2.2b shows that the Arctic has experienced more warming than the rest of the Earth, including the Antarctic, likely because the different arrangement of continents in the two hemispheres means that the oceanic and atmospheric circulation in the two hemispheres is different.

In science, no single data set is ever considered definitive, and that is particularly true of the surface thermometer record. This network of thermometers was not designed for climate monitoring. Over the years, the network has undergone many changes. Changes in the types of thermometer used, station location, the environment around the station, observing practices, and other sundry alterations all have the capacity to introduce spurious trends in the data.

For example, imagine you have a thermometer that is in a rural location in the late 19th century. Over time, a nearby city expands so that by the 1980s, the city surrounds the thermometer. Because cities tend to be warmer than nearby rural locations, this could introduce a spurious warming trend in the measurements.

Scientists know about these problems, and to the extent possible they adjust the data to take them into account. For example, the impact of a city growing up around a thermometer can be assessed by comparing the measurements from that thermometer to those from one nearby that has remained rural for the entire period. The temperature record in Figure 2.2 therefore includes adjustments in the underlying station data to account for as many of these spurious trends as possible.

Nevertheless, uncertainty in the observed warming remains, as a result of both uncertainties in the adjustments and uncertainties that cannot be adjusted for. To account for this, scientists put error bars, which are estimates of the error in their estimate, on this number. For the warming from 1906 to 2005 of 0.74 °C, the error bars are 0.18 °C. This means that the warming is likely to be between 0.56 °C and 0.92 °C, but a small chance exists that it could also lie outside this range.

Given possible problems in these data, it would be foolish to rely entirely on this single data source to determine if and how much the Earth was warming. Scientists therefore turn to other data sets to verify this result. In the rest of Section 2.1, I describe the other data sets that scientists use to build confidence in the surface thermometer data set.

2.1.2 Satellite measurements of temperature

It is possible to measure global average temperature from orbit, and the United States has been flying satellites to make that measurement since 1978. Figure 2.3 shows the time series of satellite measurements of the global monthly average temperature. These data show a general warming trend over this period of approximately 0.13 °C per decade (1.3 °C per century), which is similar to that seen in the surface thermometer record in the late 20th century.

As with all data sets, this one has its own set of problems and uncertainties. First, satellites do not measure the surface temperature; instead they measure the average temperature of the bottom 8 km of the atmosphere, from the surface to about the altitude where airliners fly. Although the temperature of this layer of the atmosphere should, in principle, generally track the surface temperature, we must be careful when directly comparing these satellite measurements of temperature to actual surface measurements.

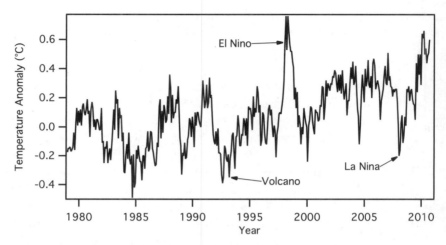

Fig. 2.3 Satellite measurements of the global monthly average temperature. The satellite measures the average temperature of a layer of the atmosphere from the surface to an altitude of approximately 8 km, about the height that airliners fly (satellite data were obtained from the University of Alabama, Huntsville; see http://vortex.nsstc.uah.edu/data/msu/t2lt/uahncdc.lt).

There are also potential problems with this data set. One example is what is known as orbit drift. Imagine that a satellite flies over a location at 2 p.m. each day and makes measurements of that location's temperature at that one time of day. Over time, the satellite's orbit drifts so that it flies over that location later and later each day. After a few years, the satellite is flying over the same location at 3 p.m. Because temperatures rise throughout the day, it is generally warmer at that location at 3 p.m. than it is at 2 p.m. Thus, the drift in the satellite's orbit would by itself introduce a warming trend, even if the temperatures were not actually changing. This artifact in the record must be identified and adjusted for.

Other issues include calibration of the satellite instruments, which were never designed to make long-term measurements, and the shortness of the satellite record (just a few decades long), both of which introduce uncertainty into the observed warming. As with the surface thermometer record, these issues are known and adjusted for, to the extent possible.

Overall, the warming trend in the satellite data is quite similar to the warming trend seen in the surface thermometer record. Thus, satellites provide strong confirmation of the reality of the warming trend. However, the question of whether the Earth is warming is so important that even more confirmation is required.

2.1.3 Ice

Because ice melts reliably at $0\,°C$, it is a dependable indicator of temperature. In particular, if the warming trend identified in the surface thermometer and satellite records is correct, then we should expect to observe the Earth's ice disappearing. In this section I show that ice is indeed disappearing, thus confirming that the Earth is warming.

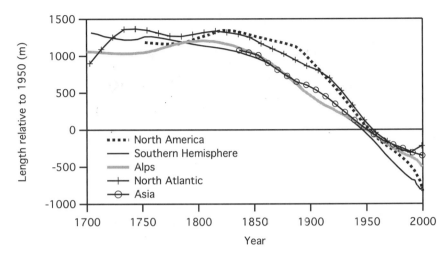

Change in mean glacier length over time, measured relative to length in 1950, for five world regions (the source is Fig. 3.2 of Dessler and Parson, 2010, which was based on Fig. 4.13 of Lemke et al., 2007).

2.1.3.1 Glaciers

Glaciers form in cold regions when snow that falls during the winter does not melt during the subsequent summer. As snow accumulates, the snow at the bottom is compacted by the weight of the overlying snow and turns into ice. Over thousands of years, this process can produce glaciers hundreds, or even thousands, of feet thick. The lengths of glaciers around the world have been measured for hundreds of years, and these records show a clear pattern of retreating glaciers. Of the 36 glaciers that were monitored between 1860 and 1900, only 1 advanced and 35 retreated. Of the 144 monitored between 1900 and 1980, 2 advanced and 142 retreated.

Figure 2.4 shows changes in average glacier length (relative to the length in 1950) for five world regions over the past few centuries. It shows that glaciers began retreating around 1800, with the shrinkage accelerating later in the 19th century. The pattern of glacier retreat is consistent worldwide; this shows that the warming we are now experiencing is truly global.

Note that decreases in precipitation or decreases in cloudiness can also cause glaciers to recede. However, the fact that glaciers are receding all over the planet means that, whatever is causing the changes, it must be global. Moreover, there is no evidence of global trends in either cloudiness or precipitation that could cause the reduction in glacier lengths. We do, however, have evidence of global trends in temperature. In this way, the recession of glaciers provides confirmation of the global warming of the climate.

2.1.3.2 Sea ice

At the cold temperatures found in polar regions, seawater freezes to form a layer of ice floating on top of the ocean, typically a few meters thick. The area covered by sea ice varies over the year, reaching a maximum in late winter and a minimum in late summer. Given the rapid warming now occurring, particularly in the Arctic, we

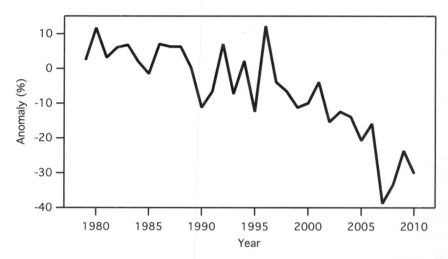

Fig. 2.5 Arctic sea-ice area anomaly in September of each year. The anomaly is the percent deviation from the 1979–2000 mean area of 7 million km² (data obtained from the National Snow and Ice Data Center; see http://nsidc.org/data/seaice_index/).

would expect to see reductions in the area covered by sea ice during the summer (during the winter, the temperatures are so low that a few degrees of warming would not have much of an effect). Figure 2.5 confirms this by showing a clear downward trend in the area covered by Arctic sea ice during September. Measurements also show that, in addition to shrinking in area, sea ice has grown thinner.

The Antarctic is a different story. The sea-ice area around that continent has remained stable since the mid-1970s. This overall pattern – large losses of sea ice in the Arctic but little loss in the Antarctic – matches the regional temperature trends in these regions. Figure 2.2b shows rapid and large warming in the Arctic but weak warming in the Antarctic. In this way, the sea-ice data strongly confirm not just the warming trend but also the global distribution of the warming.

2.1.3.3 Ice sheets

The Earth has two major ice sheets, one on Greenland in the northern hemisphere and the other on Antarctica in the southern hemisphere. Although these ice sheets are really just big glaciers, their sheer size puts them in a class by themselves. They contain the vast majority of the world's fresh water; in fact, if they melted completely, the sea level would rise approximately 100 m. These ice sheets cover millions of square kilometers of the Earth, and in places they are more than 3,000 m thick.

Figure 2.6 shows the change in mass of Greenland's ice sheet between 2002 and 2009. The measurements come from a satellite that measures the gravity of the Earth very precisely, and from that can determine changes in the mass of the ice sheets. Over this time, Greenland lost roughly 1,600 billion tons of ice – 200 billion tons or so of ice every year. Measurements from Antarctica show comparable losses for that ice sheet.

Thus, the amount of ice on the planet (glaciers, sea ice, and ice sheets) is decreasing. This is consistent with measurements of rising temperatures from the surface

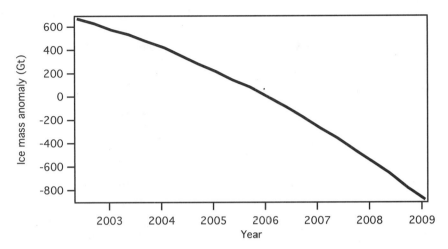

Fig. 2.6 Change in Greenland's ice mass, in billions of tons (Gt) of ice, as measured by the GRACE satellite during the period from April 2002 to February 2009. The plotted curve is a best quadratic fit to the data (adapted from Fig. 1 of Velicogna, 2009).

thermometer network and from the satellites discussed earlier. The consistency among the various data sets provides confidence that the warming they all indicate is indeed truly occurring.

2.1.4 Ocean temperatures

Much of the heat trapped by greenhouse gases goes into heating the oceans, so we can also look to see if the temperatures of the oceans are increasing. I am not talking about the surface temperature of the ocean – that is included in the surface thermometer record already described. Rather, I am talking about the temperature of the bulk of the ocean – in other words, the water temperature averaged over a significant fraction of the ocean's average depth of 4 km. Scientists determine this temperature by lowering thermometers into the ocean and measuring the temperature at various depths, and then averaging these results to come up with a single average ocean temperature over that depth.

Such measurements have been made for several decades, allowing us to determine if the temperature of the bulk of the ocean has been increasing. Figure 2.7 plots the time series of the temperature anomaly of the ocean. The ocean is indeed observed to be warming, and this provides another source of independent confirmation that the Earth is warming.

2.1.5 Sea level

Sea-level change is connected to climate change in two ways. First, as grounded ice melts,[1] the melt water runs into the ocean, increasing the total amount of water

---

[1] Grounded ice is ice that is resting on land. When it melts and the water runs into the ocean, sea level rises. This is different from floating ice. When floating ice melts, the melt water occupies exactly the same volume as was displaced by the ice (as shown by Archimedes), so there is no sea level rise. Thus, melting glaciers and ice sheets cause sea level to rise, but not melting sea ice.

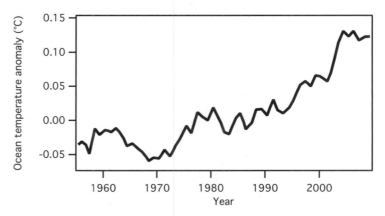

Ocean temperature anomaly in degrees Celsius for the top 700 m of the ocean. Anomalies are calculated relative to the 1957–1990 period (data obtained from http://www.nodc.noaa.gov/OC5/3M_HEAT_CONTENT/).

in the ocean and therefore the sea level. Second, like most things, water expands when it warms. We saw in Subsection 2.1.4 that the oceans are indeed increasing in temperature, so this will cause expansion of the ocean's water and a rise in sea level. Thus, a warming climate should be associated with rising seas, and measurements plotted in Figure 2.8 show that is indeed what is observed.

During the 20th century, the rate of increase in sea level was approximately 1.5 mm per year, for a total of 15 cm over the century. The few records that extend back into the 19th century suggest that the sea level rose faster in the 20th century than it did in the 19th. In recent decades, the rate of sea-level rise has increased. In the past 40 years, the increase has been roughly 1.8 mm per year, of which thermal expansion accounts for approximately one fourth (0.42 mm per year) and the melting

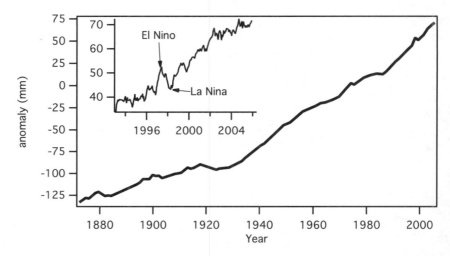

Global annual average sea-level anomaly, measured relative to the 1961–1990 average and smoothed to show decadal variations. The inset shows a close-up of unsmoothed 1993–2002 anomalies (source is Fig. 3.3 of Dessler and Parson, 2010, which was adapted from Figs. SPM.3 of IPCC, 2007a and 5.14 of Bindoff et al., 2007).

of land ice the remainder. In the most recent 10 years of data, from 1993 to 2003, the increase has been 3.1 mm per year, with more or less equal contributions from thermal expansion and melting glaciers.

I should emphasize the interconnected nature of these data sets. The observed loss of grounded ice (see Subsections 2.1.3.1 and 2.1.3.3) is adding water to the oceans. The observed warming of the ocean (see Subsection 2.1.4) causes the water to expand. From these observations, we can estimate how much the sea level should rise, which can then be compared with the actual rise in sea level. Although the calculation is difficult, and there are some aspects that are not well specified (such as the warming of the deepest part of the ocean), the observed change in sea level is consistent with the loss of ice as well as with the warming of the ocean. Such detailed comparisons further increase our confidence that the changes in each data set are accurate.

2.1.6 Putting it all together: Is today's climate changing?

The answer is an emphatic *yes!* In fact, the evidence is so strong that the IPCC calls today's warming *unequivocal* – meaning it is beyond doubt. It is worth exploring the source of high confidence in this conclusion. First, there is great consistency among the various data sets. The surface thermometer record as well as the satellite record show that temperatures are going up. The loss of ice on the Earth's surface is consistent with these increasing temperatures, as is the increase in the heat content of the ocean. Finally, the observation of increasing sea level fits with both the loss of ice and the increasing heat content of the ocean.

Importantly, these various data sets are susceptible to different kinds of errors. For example, issues such as changes in the station environment, which may affect the surface thermometer record, do not affect the satellite record. Issues such as orbit drift affect the satellite record but do not affect the surface thermometer record. And neither of these problems affects the measurements of glacier length or sea level. This means that there is no single problem or error that could push all of the temperature data in the same direction. Because of this, there is essentially no chance that enough of these sources could be wrong by far enough, and all in the same direction, that the overall conclusion that the climate is currently changing could be wrong.

Moreover, the data sources we have reviewed are just a small part of the mountain of evidence that the Earth is warming. Other corroborating evidence includes decreased northern hemisphere snow cover, thawing of Arctic permafrost, strengthening of mid-latitude westerly winds, fewer extreme cold events and more extreme hot events, increased extreme precipitation events, shorter winter ice seasons on lakes, and thousands of observed biological and ecological changes that are consistent with warming (e.g., poleward expansion of species ranges and earlier spring flowering and insect emergence). Not every single data set shows warming, but such contrary data are rare, regionally limited, and vastly outnumbered by evidence of warming.

Because of this, there should be no question in your mind that the Earth's climate is warming.

2.1.7 What is *not* evidence of climate change

It is useful at this point to also recognize what is *not* evidence of climate change. Because climate change is a shift in the statistics of the atmosphere, a single seemingly odd weather event is almost never evidence of climate change. A single extremely hot summer, for example, even if it were hotter than any other summer of the past 100 years, might occur in a stable climate. If hot summers were to begin to happen regularly, however, then that would be indicative of climate change.

It is also important to avoid drawing conclusions about the global climate from regional climate extremes. As I discussed earlier in the chapter, Figure 2.1 shows that temperatures in the Western United States and Northern Europe were much colder than average in December 2009. People living in those regions might be forgiven for wondering where global warming was, but if they concluded as a result of those regional anomalies that climate change was no longer happening, they would be wrong. In that case, other regions were hotter than average, and those compensated for the low temperatures found in the Western United States and Northern Europe.

So be careful when evaluating the evidence for and against climate change. Do not be misled by unlikely-seeming single events or by regional occurrences. Neither is indicative of a shift in the global climate.

2.2 Climate over the Earth's history

2.2.1 Paleoproxies

To put today's warming into context, it is useful to consider the Earth's entire climate history. Unfortunately, the measurements described in the previous section go back at most a few centuries, so other data sets are needed if we wish to look back any further. Such data sets are known as *paleoproxies*, which are long-lived chemical or biological systems that have the climate imprinted on them. In this way, we can make measurements *today* that tell us what the climate was like *in the past*.

For example, the ice in a glacier or ice sheet provides useful climate data dating back to the time when the snow fell. Remember that glaciers and ice sheets form when snow accumulates from one year to the next and is converted to ice by the weight of the overlying snow. The chemical composition of the ice holds important information about the air temperature around the glacier when the snow fell, as do variations in the size and orientation of the ice crystals. Small air bubbles trapped during the formation of glacial ice preserve a snapshot of the chemical composition of the atmosphere at that moment. In addition, the dust trapped in the ice gives information about prevailing wind speed and direction, and about how wet or dry the regional climate was when the ice formed, because more dust blows around during droughts. Finally, because sulfur is one of the main effluents of volcanoes, measurements of sulfur (usually in the form of sulfuric acid) in glacial ice can show whether there was a major volcanic eruption around the time the ice formed.

To obtain all of this information, *ice cores* are used. Ice cores are obtained by drilling down through the entire depth of a glacier or ice sheet with a hollow drill bit and extracting a thin cylinder of ice a few inches in diameter. Reconstructing past climate information from an ice core then requires two steps. First, the age of each ice layer must be determined from its depth inside the glacier. The deeper down the ice was obtained, the older the ice is and the further back the time for which it provides climate information. Much effort has been spent connecting a particular chunk of ice to an exact time, because the rate of ice accumulation varies over time and because ice inside a glacier can compress and flow under the great weight of the ice above. Second, the characteristics actually observed, such as the abundance of chemicals in the ice, must be translated into the climatic characteristics of interest, such as temperature. Ice cores from the thickest, oldest ice sheets in Antarctica have provided climate reconstructions dating back an amazing 800,000 years.

Obviously, ice cores only provide climatic information in regions and over time periods that are cold enough for permanent ice to exist. But there are other paleoproxies that provide data in other regions and over other time periods. For example, trees also store climate information in their tree rings. Tree growth follows an annual cycle, which is imprinted in the rings in their trunks. As trees grow rapidly in the spring, they produce light-colored wood; as their growth slows in the fall, they produce dark wood. Because trees grow more, and produce wider rings, in warm years, the width of each ring gives information about climate conditions around that tree in that year. By looking at the rings of a tree, scientists can obtain an estimate of the local climate around the tree for each year during which the tree was alive.

Climate data from tree rings are only available for a fraction of the Earth's surface. They are obviously not available for oceans, or for desert or mountainous areas where no trees grow. They are also not available in the tropics, where the weaker seasonal cycle causes trees to grow year round; these trees do not produce rings. Finally, tree rings only reveal information about the climate as far back as trees are available. This means that tree rings tell us about the climate over the past 1,000 years.

Ocean sediments, which accumulate at the bottom of the ocean every year, also contain information about climate conditions at the time they were deposited. The most important source of information in sediments comes from the skeletons of tiny marine organisms. The relative abundance of species that thrive in warmer versus colder waters gives information about surface water temperature. The chemical composition of the skeletons and variations in the size and shape of particular species provide additional clues. In total, ocean sediments can provide information about water temperature, salinity, dissolved oxygen, atmospheric carbon dioxide, nearby continental precipitation, the strength and direction of the prevailing winds, and nutrient availability; this information goes back tens of millions of years.

Putting all of these paleoproxies together gives us a reasonably complete picture of the global climate going back many millions of years, with some information about the climate going back billions of years.

2.2.2 The Earth's long-term climate record

Although many of details of the climate during the first 97% of the Earth's history are unknown, there are a few things that we do know. To begin with, the oldest sedimentary rocks on the planet are nearly 4 billion years old. Because sedimentary rocks generally form in the presence of liquid water, their existence shows that the

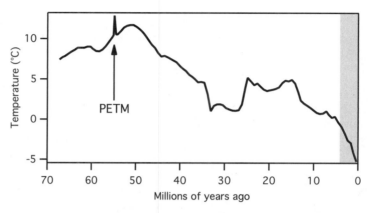

Fig. 2.9 Reconstructed temperature of the polar regions over the past 70 million years. The sharp temperature spike 55 million years ago represents the PETM. The gray bar on the right shows the past 4 million years, which are expanded in Figure 2.10. Because this time series is from ocean isotopes, it is sensitive both to temperature and to the total volume of ice on land. Starting roughly 35 million years ago, some of the variation here comes from changes in land ice volume rather than temperature. The overall trend, however, mostly represents changes in temperature (the source is Fig. 3.8 of Dessler and Parson, 2010, which was adapted from Fig. 2 of Zachos et al., 2001).

Earth has been warm enough over most of its history that water on the planet has remained mostly in the liquid phase.

But while the Earth has generally been warm, there is also evidence of intervals of widespread ice cover (known as a glaciation). The evidence comes in the form of marks on rocks, such as abraded rock surfaces and other geologic formations that are formed when giant ice sheets flow over rocks. There is evidence that, approximately 700 million years ago, the Earth was covered by ice from the poles to near the equator – a climate configuration now referred to as *snowball Earth*. There was also significant planetary glaciation roughly 300 million years ago. Over most of the past several hundred million years, however, the planet has been relatively warm.

Figure 2.9 shows a reconstruction derived from ocean sediments of polar temperatures over the past 70 million years. The warmest temperatures in this record occurred approximately 50 million years ago – 15 million years or so after the extinction of the dinosaurs – in a period called the *Eocene Climatic Optimum*. During that time, the planet was far warmer than it is today. Forests covered the Earth from pole to pole, and plants that cannot tolerate even occasional freezing lived in the Arctic, along with animals such as alligators that today live only in the tropics. Since that time, the Earth has experienced significant long-term cooling. Clearly, humans had nothing to do with either the warmth of the Eocene or the cooling since then; I will talk more about the causes of these climate variations in Chapter 7.

Figure 2.9 also shows the Paleocene–Eocene Thermal Maximum, more commonly known as the PETM. This was an abrupt, brief period of warming that occurred 55 million years ago or so, at the temporal boundary between the Paleocene and Eocene epochs, in response to a massive release of greenhouse gases. Many people

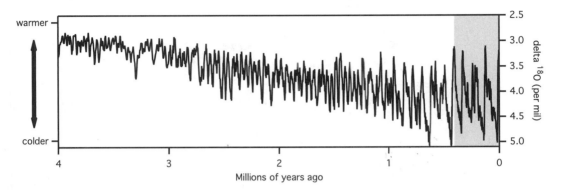

Fig. 2.10 Measurement of global average relative temperature over the past 4 million years. The vertical axis measures the relative abundance of oxygen-18, the heavy isotope of oxygen that is a proxy for temperature, in ocean sediment cores. The temperature difference between the top and bottom of the graph is roughly 10 °C. The gray bar on the right shows the past 410,000 years, which are expanded in Figure 2.11 (the source is Fig. 3.9 of Dessler and Parson, 2010, which was based on the analysis of Lisiecki and Raymo, 2005).

view this as a good analog to the warming we are now experiencing. I will discuss this event in more detail in Chapter 7.

Figure 2.10 zooms in to show global temperature variations over the past 4 million years. Like the 70-million year record in Figure 2.9, this record also shows a general cooling trend. This record, however, covers a more recent time, so we can see fine-scale details that are not visible in the longer-term record. For example, starting roughly 3 million years ago, around the time that large ice sheets first appeared in the northern hemisphere, large oscillations between warmer and cooler periods suddenly appear in the record. During the cool periods, called *ice ages*, the ice sheets expanded to cover large parts of the northern hemisphere. During the warm periods between the ice ages, called *interglacials*, the ice sheets contracted. From approximately 2.5 million to 1 million years ago, ice ages occurred every 41,000 years. Since then, for reasons that are not well understood, the frequency of ice ages shifted to every 100,000 years and the magnitude of the ice-age cycles increased.

Figure 2.11 zooms in again, showing a record constructed from ice cores of temperature and carbon dioxide levels for the Antarctic region over the past 410,000 years. This record shows still more finely grained detail than the longer data sets just described, including the shape of ice-age cycles. The cooling into an ice age is slow, taking several tens of thousands of years, whereas the warming at the end of an ice age occurs faster, in approximately 10,000 years. The interglacials are relatively short, lasting 10,000–30,000 years, whereas the ice ages last 100,000 years or so. The last ice age ended roughly 10,000 years ago, and since then the Earth has been enjoying a pleasant interglacial.

Note that ice ages are only 5–8 °C colder than today, which is a seemingly small difference if we consider that the Earth is essentially a different planet during an ice age, with glaciers several thousand feet thick covering much of North America, a sea level that is 100 m lower than that today, and all of the other accompanying changes in the world's environment and ecosystems. Also note that atmospheric carbon dioxide, which can also be estimated from the ice core, varies closely with

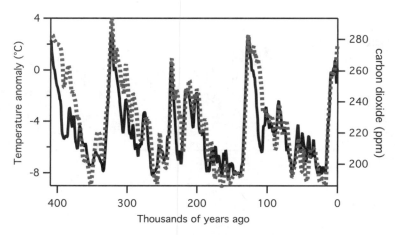

Fig. 2.11   Temperature anomaly of the southern polar region (solid line) over the past 410,000 years, relative to today's temperature, constructed from an Antarctic ice core. Carbon dioxide (dotted line) is from air bubbles trapped in the ice (the source is Fig. 3.10 of Dessler and Parson, 2010, which was adapted from Petit et al., 1999).

atmospheric temperature over these ice age cycles. I will discuss the implications of this relationship in detail in Chapter 7.

Finally, Figure 2.12 zooms in one last time to show average northern hemisphere temperature over the past 1,000 years, based on multiple proxies and modern records. Once again, this figure shows short time-scale temperature variations that are not visible in the graphs covering longer time spans. The various estimates differ, particularly before Year 1500 or so, but all show a similar pattern. Temperatures were warm 1,000 years ago, during a period known as the *Medieval Warm Period*. There

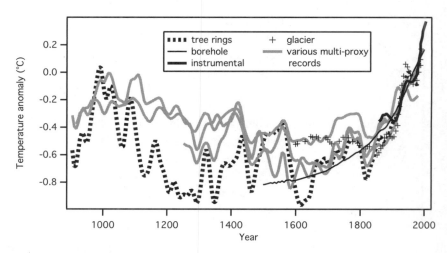

Fig. 2.12   Average northern hemisphere temperature anomaly over the past 1,000 years, based on multiple proxy records and the modern surface thermometer record. Anomalies are calculated relative to the 1961–1990 average (the source is Fig. 3.11 of Dessler and Parson, 2010, which was adapted from Fig. S-1 of the National Research Council, 2006).

were then several centuries of gradual cooling, bottoming out in a period 200–300 years ago known as the *Little Ice Age*, followed by faster warming since the 19th century.

This vast and growing body of information about the Earth's past climate places the climate of the past century in context. We can say with high confidence that the past few decades of the 20th century are warmer than any comparable period during the past 400 years, and possibly even warmer than the peak of the Medieval Warm Period, around 1,000 years ago. We can also say with high confidence that, when we consider geologic time scales (millions or billions of years), the Earth has been both far warmer and far cooler than it is today. This latter fact should not be taken to suggest that today's warming must therefore be natural. As I will discuss in Chapter 7, there is strong evidence to suggest that today's warming is primarily due to human activities.

We can also say that the warming of the past few decades has been rapid. For example, the warming over the past century (approximately $0.74\,^\circ$C in 100 years) is occurring 20 or so times faster than the average rate of warming since the height of the last ice age (roughly 5–8 $^\circ$C in 20,000 years). This means that projections of warming of several degrees Celsius over the next century (which I discuss in Chapter 8) would be both large and exceptionally rapid, and these should compel our attention.

The challenge for the scientific community is to come up with a theory that explains all of the variations in the climate record, from snowball Earth 700 million years ago to the rapid warming of the past few decades. In the next few chapters we will learn about the fundamental physics that governs our climate, and then in Chapter 7 we will put it all together to show that most of the warming of the past few decades can be attributed to human activity.

2.3 Chapter summary

- The most well-studied and reliable source of temperature data for the past century is the surface thermometer network. It shows a global annual average warming of $0.74\,^\circ$C over the course of the 20th century, with an uncertainty of $0.18\,^\circ$C.
- Scientists have a large number of independent measurements with which to confirm the warming seen by the surface thermometer network. These include satellite measurements of temperature, measurements of the amount of ice on the planet, ocean heat measurements, and sea-level measurements. All of these data confirm the warming seen in the surface thermometer data.
- Because of the overwhelming evidence supporting it, the scientific community has concluded that the observed warming of the climate system is beyond doubt – the IPCC says it is unequivocal. Furthermore, scientists have concluded that the previous decade is very likely the hottest of the past 400 years.
- Looking back further in time, we see that the Earth's climate has varied widely over its 4.5-billion-year history. The geologic record shows that the climate has been both much warmer and much cooler than today's climate.

- Over the past few million years, the Earth has oscillated between ice ages and warmer interglacial periods. Ice ages are 5–8 °C cooler than the interglacials. The Earth is currently in an interglacial.

Additional reading

T. R. Karl, S. J. Hassol, C. D. Miller, and W. L. Murray (eds.), *Temperature Trends in the Lower Atmosphere: Steps for Understanding and Reconciling Differences* (Washington, DC: U.S. Climate Change Science Program, 2006). This assessment by the U.S. Climate Change Science Program describes the numerous details of a calculation of the global average temperature. The trends in several important data sets are compared, and it concludes that the trends are all generally consistent, although some discrepancies do exist (download at http://www.climatescience.gov/Library/sap/sap1–1/default.php).

K. E. Trenberth et al., "Observations: Surface and Atmospheric Climate Change," in S. Solomon, D. Qin, M. Manning, et al. (eds.), *Climate Change 2007: The Physical Science Basis*. Contribution of Working Group I to the Fourth Assessment Report of the Intergovernmental Panel on Climate Change (Cambridge: Cambridge University Press, 2007). This is the IPCC's latest evaluation of the scientific evidence that the climate is changing. Although it is not necessarily easy for nonscientists to read, this chapter summarizes in great detail much of the definitive evidence that the climate is indeed changing (download at http://www.ipcc.ch).

Terms

Eocene Climatic Optimum
Ice ages
Ice core
Interglacials
Little Ice Age
Medieval Warm Period
Paleoproxies
Snowball Earth
Temperature anomaly

Problems

1. How much has the Earth warmed over the past 100 years? Provide the answer in both degrees Celsius and degrees Fahrenheit.

2. If you found out that the satellite data were unreliable because of a previously unknown error, would that change your opinion about whether the Earth is currently warming? Why or why not?

3. A reporter asks you to explain why scientists are so confident that the Earth has undergone a general warming over the past few decades. Knowing that reporters hate long answers, write an answer that takes 60 seconds or less to deliver.

4. List the evidence that supports the contention that the Earth is currently warming. Is there any evidence that goes against this conclusion?

5. What is a temperature anomaly? Why are temperature anomalies typically used in global temperature calculations?

6. Download the annual and global average temperature data from the NASA GISS (Google the term *GISTEMP*) and reproduce Figure 2.2. Calculate your own trends for the past 30 years and for the past 100 years.

7. Why do we turn to paleoproxy measurements to infer the temperature of millions of years ago?

8. Go to http://cdiac.ornl.gov/epubs/ndp/ushcn/access.html and plot up the monthly temperature for the past 100 years for the station nearest your hometown. Does this look like the global average time series in Figure 2.2? Should it?

9. A global warming advocate tells you that the Earth is now warmer than it has ever been. Is that correct?

Radiation and energy balance

The Earth's climate is a complex system. Luckily, we can still understand a lot about the climate even if we do not have an advanced degree in physics. In this chapter, I introduce the important physics that we need to know to understand the climate; in Chapter 4 I will use this physics to create a simple model of our climate.

3.1 Temperature and energy

Before we get into the physics of radiation and energy balance, it is useful to talk about the concept of *energy*. To a physicist, energy is the capacity to do work – such as lifting a weight, turning a wheel, or compressing a spring. The unit of energy most frequently used in physics is the *joule*, abbreviated as the letter *J*. Energy is sometimes expressed in units of calories, or cal; 1 cal = 4.18 J. A food calorie is actually 1,000 cal, also called a kilocalorie or large calorie (1 kcal = 1,000 cal or 4,184 J). If you go to Europe, the nutritional label on food packaging has the energy content of the food marked in joules rather than food calories. Thus, a bag of Cheetos with 300 kcal or food calories would instead be labeled as containing 1.3 MJ or megajoules. I prefer food labeled in food calories because 1.3 megajoules sounds more fattening than 300 calories!

Energy often moves from one place to another. The rate at which energy is moving is referred to as *power*. It is usually expressed in *watts*, abbreviated as the letter *W*. One watt is equal to one joule per second – that is, 1 W = 1 J/s – so a 60-W light bulb consumes 60 J of energy every second.

An analogy may help to illuminate the difference between power and energy. A gallon is a measure of a quantity of water, just like a joule is a measure of a quantity of energy. The rate at which water flows through the pipe is measured in, say, gallons per minute. An analogous flow of energy is expressed in units of watts (joules per second). As you will see in this chapter and the next, climate is all about energy flows, so calculations of climate physics generally focus on power.

An example: How much power does it take to run a human body?

A typical human consumes approximately 2,000 food calories (equal to 2,000 kcal) per day. We know that 2,000 kcal × 4,184 J/kcal = 8,368,000 J = 8.37 MJ. One day has 86,400 s (or seconds) in it, so dividing 8,368,000 J by 86,400 s, we get 97 J/s = 97 W. Thus, the typical human requires roughly 100 W to power his or her body – about the same power required to run a typical light bulb. One horsepower (1 hp) is approximately 740 W, so another way to think about this is that it takes one seventh of a horsepower or so (0.14 hp) to run your body.

The *internal energy* of an object refers to how fast the atoms and molecules in the object are moving. In a cup of water, for example, if the water molecules are moving slowly then the cup has less internal energy than another cup in which the molecules are moving rapidly. In a solid, the movements of the atoms are approximately fixed in space by intermolecular forces – that is why it is a solid. The atoms, however, can still move small distances around their fixed position. The faster these atoms move about their fixed position, the more internal energy the object has.

This brings us to a concept that most people are familiar with: *temperature*. Temperature is a measure of the internal energy of an object. As an object's internal energy increases – and the molecules of the object speed up – the temperature of the object also increases. Thus, if you have two cups of water, one hot and the other cold, you can conclude that the water molecules in the hot cup are moving faster than molecules in the cup of cold water.

In Chapter 1, I introduced the Celsius temperature scale, which is frequently used by scientists. There is another temperature that is even more favored by physicists, and it is called the Kelvin scale. The temperature in degrees Kelvin is equal to the temperature in degrees Celsius plus 273.15 (K = C + 273.15). Thus, the freezing temperature, 0 °C, is equal to 273.15 K, whereas the boiling temperature, 100 °C, is equal to 373.15 K. "Room temperature" is 22 °C or so, which is approximately 295 K. Most temperatures found in the Earth's atmosphere are between 200 K and 300 K, and the average surface temperature of the Earth (today, at least) is roughly 288 K.

Physicists prefer the Kelvin scale because temperature expressed in degrees Kelvin is proportional to internal energy. Thus, if the temperature doubles from 200 K to 400 K, then the internal energy of the object also doubles. If the internal energy of an object increases by 10%, then the temperature expressed in degrees Kelvin also increases by 10%. This means that an object at 0 K, also called absolute zero, has an internal energy of zero – meaning that the constituent molecules are not moving. Because of these important qualities, the physics equations introduced in this chapter and the next require energy to be expressed in degrees Kelvin. Thus, I will use Kelvin temperatures in my climate calculations, whereas I will primarily use Celsius temperatures descriptively.

3.2 Electromagnetic radiation

It has been recognized for a long time that the warmth of our climate is provided by the Sun. However, the Sun sits 150 million km away from the Earth, with the vacuum of space in between. How does energy from the Sun reach the Earth?

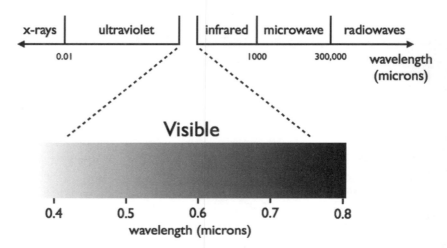

Fig. 3.1 The electromagnetic spectrum. Note that the visible part makes up only a minor part of this spectrum. (See Color Plate 3.1.)

Energy is transported from the Sun to the Earth by what is known as *electromagnetic radiation*.[1] Electromagnetic radiation includes visible light, like that put out by your desk lamp or the Sun; X rays that allow us to detect broken bones; microwaves that cook your dinner; and radio-frequency waves that bring calls to your cell phone and WiFi to your computer.

Electromagnetic radiation emanating from a flashlight, a lamp, your WiFi router, or the Sun is really a stream of *photons*, small discrete packages of energy.[2] As photons travel from Point *A* to Point *B* – such as from the Sun to the Earth – each one carries a small amount of energy, and this is how energy is transported from the Sun to the Earth.

Photons have a characteristic size, referred to as the *wavelength*, which determines how the photons interact with the world. Photons with wavelengths of between 0.3 and 0.8 micrometers, abbreviated as μm (a *micrometer* or micron is a millionth of a meter; a human hair is 100 μm or so in diameter), can be seen with the human eye – so we refer to these photons as *visible*. Within the visible range, the different wavelengths appear to the human eye as different colors (Figure 3.1). Humans see photons with wavelengths near 0.4 μm as blue, 0.6 μm as yellow, and 0.8 μm as red.

Photons with longer wavelengths, from 0.8 to 1,000 μm, are termed *infrared* – from the Latin for "below red" – because they are beyond the red end of the visible spectrum. Despite being invisible to humans, these photons play an important role in both the Earth's climate and in our everyday lives. Photons with wavelengths

---

[1] When people hear the word *radiation*, they often think of nuclear radiation. Such radiation has very high energies because it originates from changes in atomic nuclei, and as a result this radiation can cause cancer and other medical problems. Electromagnetic radiation discussed here generally originates from changes in the atoms' electrons or from changes in the molecule's rotational or vibrational state, and therefore has far less energy – so it is generally not a health hazard. This is good, because you are surrounded by electromagnetic radiation right now.

[2] Electromagnetic radiation also behaves like a wave, but for this problem it is easier to think of it as a particle.

just below the human detection limit of 0.3 μm are called *ultraviolet* because their wavelength is beyond the violet end of the visible limit.

Photons with wavelengths between 1,000 μm (1 mm) and 0.3 m are termed *microwaves*, and photons in this wavelength range are used in many familiar applications, from cooking to radar. Wavelengths bigger than about 0.3 m are radio-frequency waves, and they are used, as the name implies, in radio. The entire electromagnetic spectrum is diagrammed in Figure 3.1.

The wavelength of the photon determines its physical properties. For example, visible and infrared photons cannot go through walls, but radio-frequency photons can. The human eye can detect visible photons, but not infrared or microwave photons. When you get a full body scan at the airport, the machine is using either X rays or microwaves – both wavelengths can go through clothes but are stopped by denser materials such as flesh or a bomb. Finally, the atmosphere is transparent to visible photons but less transparent to infrared photons; this fact has enormous implications for our climate and will be discussed at length in Chapter 4.

3.3 Blackbody radiation

We know that both the Sun and the lamp on your desk are emitting photons. After all, you can see the visible photons that they are emitting. They are not, however, the only things around you that are emitting photons. In fact, *everything around you* is emitting photons all of the time. So right now, you're emitting photons, as are the walls of the room you're sitting in, your desk, your dog, this book. Everything.

If everything is emitting photons, then why doesn't everything glow like a light bulb? The reason is that an object emits photons with a wavelength determined by the object's temperature. Figure 3.2 plots *emissions spectra* for idealized objects called *blackbodies* at three temperatures. An emissions spectrum is the amount of power carried away from an object by the photons at each wavelength.

As shown in Figure 3.2a, photons emitted by objects at room temperature, approximately 300 K, almost exclusively have wavelengths greater than 4 μm or so. These wavelengths are outside the range that is visible to humans (indicated by the gray shading in the figure). Thus, all room-temperature objects are emitting photons, but you cannot see the photons because they fall outside the visible range. This is, in fact, the origin of the term *blackbody*. At room temperature, the object appears black because the photons emitted by these objects are invisible to humans.

Figure 3.2a shows that the peak of the emissions spectrum for a 300-K object occurs near 10 μm and most of the energy being emitted by a room-temperature object occurs through the emission of photons near this wavelength. It turns out that there is a simple relation between the temperature of an object and the peak of the object's emission spectrum. This relation is known as Wien's displacement law:

$$\lambda_{max} = \frac{3000}{T} \qquad (3.1)$$

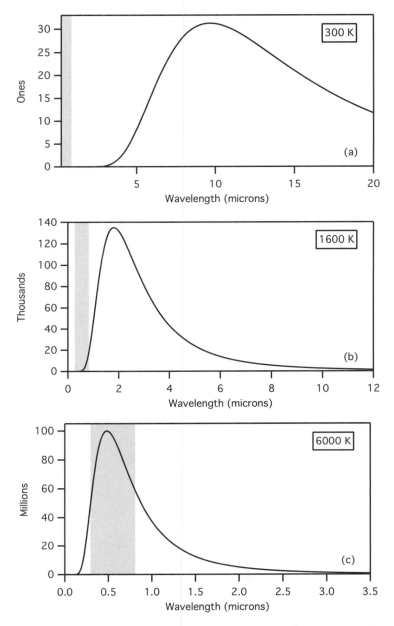

Fig. 3.2 Power emitted at different wavelengths from objects (with surface area of 1 m^2) at three temperatures: (a) 300 K, (b) 1600 K, and (c) 6000 K. The vertical axes are in units of 1 W/μm, 1,000 W/μm, and 1 MW/μm of wavelength range, respectively. Gray bars show the wavelength range visible to human eyes.

Here λ_{max} is the wavelength of the peak of the emission spectrum and T is the temperature of the object. If we put 300 K into Equation 3.1, we get 10 μm, which is in good agreement with Figure 3.2. Note that T must be in degrees Kelvin and λ_{max} must be in micrometers in this equation. Had I used the temperature in degrees Celsius, for example, I would have calculated $\lambda_{\mathrm{max}} = 3{,}000/24 = 125$ μm, which would be incredibly wrong.

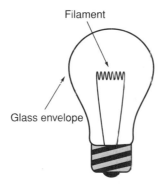

Filament

Glass envelope

Fig. 3.3 A schematic of a typical incandescent light bulb.

Wien's displacement law also tells us that, as an object heats up, the peak of its emission spectrum moves to shorter wavelengths; in other words, λ_{max} becomes smaller. Figures 3.2b and 3.2c show that a 1600-K object has $\lambda_{max} = 1.9$ μm and a 6000-K object has $\lambda_{max} = 0.5$ μm.

It is also clear from Figure 3.2 that objects do not just emit photons at λ_{max}; they also emit them over a range of wavelengths around λ_{max}. So while $\lambda_{max} = 1.9$ μm for the 1600-K object, the object emits photons over a range of wavelengths from 0.7 to 10 μm. Because a small fraction of the photons emitted by this object have wavelengths smaller than 0.8 μm, which are visible and lie at the red end of the visible spectrum, humans will perceive a 1600-K object to have slight reddish glow to the object. In other words, this object is glowing "red hot." Blacksmiths use this fact to determine when a piece of metal has reached the appropriate temperature, and the necessity of seeing a faint glow from an object is one reason that blacksmiths tend to work in dim, low-light conditions.

For the 6000-K object, most of the photons emitted fall within the visible range. Our Sun is, to a good approximation, a 6000-K blackbody, and the distribution of photons from the Sun is closely approximated by the emissions spectra in Figure 3.2c. Because being able to see confers a strong advantage in surviving, it is no surprise that the eyes of humans and other animals have evolved to see this range of wavelengths. In fact, the human eye is maximally sensitive to light with a wavelength near 0.5 μm, which is the λ_{max} for a 6000-K blackbody. The chlorophyll molecule, the key component of photosynthesis, strongly absorbs photons in the visible range, showing that plants have also evolved to take advantage of photons emitted by the Sun.

Finally, if the photons emitted by room-temperature objects are not visible to our eyes, how can we see room-temperature objects, such as this page? What you see when you look at a room-temperature object are visible photons (emitted by the Sun or a light bulb or some other sight source) that have bounced off the object.

An everyday object that uses a lot of the concepts that we have discussed in this chapter is the humble incandescent light bulb. An incandescent light bulb consists of a glass envelope containing a small filament made of a metal, such as tungsten. When the light bulb is turned on, electricity flows through the filament, heating it to around 3000 K (Figure 3.3).

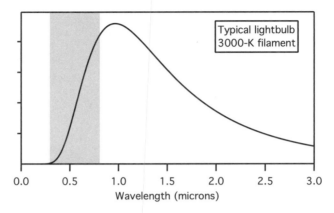

Fig. 3.4 Emissions spectrum for a 3000-K blackbody, a typical filament temperature for an incandescent light bulb. The numbers on the y-axis are omitted.

Figure 3.4 shows the wavelength distribution of photons emitted by a 3000-K blackbody. As the figure shows, the filament is hot enough that some of the photons emitted are visible – so humans will see the light bulb glowing and you can use it to light your room. However, nearly 85% of the photons emitted have wavelengths too long for the human eye to detect. These photons are basically wasted, and this makes incandescent light bulbs extremely inefficient as light sources.

One way for a light bulb to produce a higher fraction of visible photons – and therefore be more efficient – is to run the filament at a higher temperature. As described by Equation 3.1, this shifts the distribution of emitted photons to shorter wavelengths, thereby making a greater fraction of them visible to humans. The problem is that, as the temperature of the filament increases, the bulbs tend to burn out quickly. To get around this problem, the light bulb could be filled with halogen gas instead of the nitrogen and argon found in most incandescent bulbs. Because of chemical reactions between the halogen gas and the filament, these so-called halogen light bulbs can be run at temperatures several hundred degrees hotter than a regular incandescent bulb. This means that halogen light bulbs put out more photons in the visible range, making them more efficient than regular incandescent bulbs. Unfortunately, because the filament is run so hot, the light bulb itself also gets extremely hot, creating a fire and burn hazard.

As Figure 3.2 shows, the optimal temperature for the filament would be about the temperature of our Sun, nearly 6000 K, which provides the best overlap between blackbody emission and the human visual range. Unfortunately, it is impossible to run any kind of incandescent bulb at such temperatures because the filament would immediately vaporize and the bulb would be destroyed.

A better way to obtain high efficiency is to change the technology. The compact fluorescent light bulb, or CFL, uses a different technology (which I will not discuss here) to emit most of the bulb's photons in the visible wavelength range. Because of this, you get about the same amount of light out of a compact fluorescent as you do out of an incandescent bulb that consumes four to six times as much power. For example, a 12-W CFL will produce the same amount of light as a 60-W incandescent light bulb. In an effort to make the country more energy efficient, the U.S. Congress passed a law in 2007 phasing out standard incandescent bulbs in the United States by

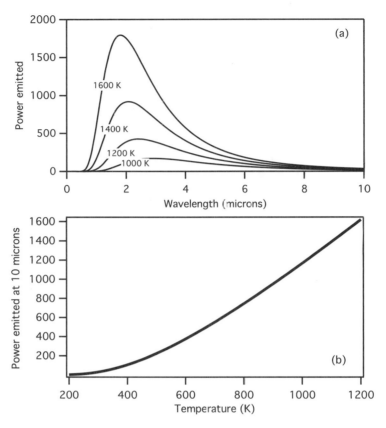

Fig. 3.5 Plots of (a) the distribution of power emitted by a blackbody at four different temperatures (1600, 1400, 1200, and 1000 K) and (b) energy emitted by a blackbody at 10 μm as a function of temperature. Plotted quantities are in $(W/m^2)/\mu m$.

2014. After that time, only energy-efficient CFLs and other new technologies will be available in stores.

Not only does the wavelength of emission change as an object's temperature changes, but the total power emitted also increases with temperature. We could see this in Figure 3.2, but it is more explicitly shown in Figure 3.5a, which shows four different blackbody-emission curves on a single plot. The plot shows that warmer objects emit more power than cooler objects at all wavelengths.

For a different view of this, Figure 3.5b plots the power emitted at 10 μm as a function of a blackbody's temperature. It is apparent that, as the temperature of the object increases, so does the power emitted at this wavelength. Infrared thermometers, which you can buy in any store, measure the emitted power at a single wavelength and then use a relation like the one in Figure 3.5b to estimate the temperature. Astronomers also use this principle to infer the temperature of distant stars and planets.

Figure 3.6 shows an image of a friendly dog in the infrared. To construct this image, the temperature is determined by measuring the power emitted at a particular wavelength and converting this to temperature. Bright colors in the image indicate warm temperatures and dark colors indicate cool temperatures. Like humans, dogs are mammals and their body temperature is around 38 °C. Fur is an insulator, however, so fur-covered regions of the dog are closer to room temperature than to body

36.0
34.1
31.9
30.0
27.9
25.9
23.8
21.2

Fig. 3.6 Photo of Bailey the dog in the infrared, with colors assigned to different temperatures. Photo courtesy of New Mexico Tech Department of Physics. (See Color Plate 3.6.)

temperature. The parts of the dog that are not fur covered, however, such as the eyes, mouth, and the inside of the ears, are close to the dog's internal temperature. Note also the dog's cold nose.

As I already mentioned, the total power emitted by a blackbody increases with temperature. There is, in fact, a simple relation, known as the Stefan–Boltzmann equation, between the total power radiated by a blackbody and temperature:

$$P/a = \sigma T^4 \tag{3.2}$$

Note that P/a is the power emitted by a blackbody per unit of surface area, with units of watts per square meter; σ is the Stefan–Boltzmann constant, with $\sigma = 5.67 \times 10^{-8}$ (W/m^2)/K^4; and T is the temperature of the object in degrees Kelvin. If you multiply P/a by the surface area a of the object (in square meters), then you get the total power emitted by a blackbody, in watts.

An example: How fast is a room-temperature basketball losing energy by the emission of photons?

At room temperature, a blackbody is emitting $\sigma(300 \text{ K})^4 = 460$ W/m^2. A basketball typically has a radius of 5 in. $= 0.13$ m, so its surface area is therefore $4\pi(0.13 \text{ m})^2 = 0.2$ m^2. Thus, the total rate of energy loss from a room-temperature basketball as a result of blackbody photon emission is 460 W/m$^2 \times 0.2$ m^2, or 92 W. Thus, the amount of power being radiated is about the same as a typical light bulb. Of course, you cannot light a room with a basketball because the photons emitted by a basketball are outside the range that humans can see.

3.4 Energy balance

One of the cornerstones of modern physics is the first law of thermodynamics, which basically says that *energy is conserved*. What this means is that if some object loses some energy, then some other object must gain that same amount of energy. Furthermore, because photons are just little packets of energy, the first law tells us that when an object emits a photon, the emitting object's internal energy must decrease. And because temperature is a measure of internal energy, the emission of a photon therefore causes the temperature of an object to decrease. Similarly, if a photon hits an object and is absorbed, then the energy of the photon is transferred to the object's internal energy and the object's temperature will increase.

An example: conservation of money

A good analogy for energy balance is money balance. If you gain 1 dollar, then someone else must be 1 dollar poorer – because money, like energy, cannot be created or destroyed.[3]

Consider, for example, a checking account. Money, such as your paycheck or a birthday check from your grandmother, is periodically deposited into the account. At the same time, money is withdrawn, to pay for things such as rent or a cell phone bill. The change in your bank balance is equal to the difference between the total deposits (money in) and total withdrawals (money out). In equation form, we write this as follows: Change in balance = money in − money out. If money in exceeds money out, that is, your deposits exceed your withdrawals, then the change in balance is positive and your balance increases. If money out exceeds money in, then the change in balance is negative and your balance decreases. If money in and money out are equal, the change in balance is zero and your balance is unchanged. This is basically the calculation we do when we balance our checkbooks.

If the energy flowing into an object (energy in) exceeds the energy flowing out (energy out), then the internal energy (and temperature) of the object increases. Written mathematically, this is as follows:

$$\text{Change in temperature} \propto \text{Change in internal energy} = \text{energy in} - \text{energy out}$$

Here the symbol $\propto$ means "is proportional to." Note the special case in which energy in and energy out are equal, in which case the internal energy and temperature are unchanged. We call this situation *equilibrium*.

A good example that draws many of the concepts in this chapter together is the oven in your home. Most people, if asked how an oven cooks, would answer, "because it's hot inside." However, you may be surprised that the physics is subtler than you realize. Ovens do not cook because the air in the oven is hot – air is a terrible conductor of heat. Rather, ovens cook by infrared radiation.

When an electric oven is turned on, electricity runs through a heating element. The element heats up, eventually reaching temperatures high enough that it radiates in the visible range, glowing a dark orange. At this point, the element is radiating an enormous amount of power, typically several thousand watts.

---

[3] This rule does not apply to national governments, which can print money.

The photons emitted by the heating element are absorbed by the walls of the oven, heating them. When the walls reach a predetermined temperature, typically 350–450 °F (450–500 K), the oven is "preheated," and the cook puts the food, say a turkey, into the oven. Let's assume the turkey came out of the refrigerator and has a temperature of 3 °C or 276 K. At this temperature, the turkey is radiating 330 W/m². If the turkey has a surface area of 0.1 m², then the total power radiated by the turkey is 33 W.

The turkey is also absorbing photons from the oven's hot walls. The oven walls, at 375 °F (465 K), are radiating 2,650 W/m². The total surface area of the oven's six walls is approximately 1.3 m², so the total power radiated by the oven's wall is roughly 3500 W. Most of the energy radiated by the oven's walls misses the turkey in the middle and hits the other walls, and only a fraction of photons emitted by the walls hits the turkey. It turns out that the turkey absorbs photons emitted by an area of the walls equal to the surface area of the turkey, 0.1 m². Given that the walls emit 2650 W/m², that means that the turkey is absorbing 265 W of power.

Because the turkey is emitting 33 W but absorbing 265 W, the internal energy of the turkey is increasing and it is therefore warming. Eventually, the turkey reaches the temperature when it is considered "done" and the cook removes it from the oven. That's how a conventional oven cooks.

While the turkey is absorbing energy from the walls, by conservation of energy the walls must be losing energy and cooling down. As a result, the heating element of the oven has to turn on periodically to maintain the wall temperature at 375 °F. This occasional cycling back on of the heating element is familiar to any cook.

A microwave oven also cooks food by bombarding food with photons. However, instead of bombarding the food with infrared photons, a microwave oven bombards the food with microwave photons, which have longer wavelengths. For reasons that we will not go into here, microwave ovens cook faster because they are able to deliver higher rates of power to the food than a conventional oven can. In the example here, the oven is delivering 265 W of power to the turkey. By using microwaves, however, the oven is able to deliver 5 to 10 times that amount. The net result is that the food is heated more rapidly in a microwave oven than it is in a conventional oven.

I hope that you have a sense of the importance of the physics we have discussed in this chapter – it has a profound impact on your life and the world around you. It also plays a key role in climate. In Chapter 4, we will use the physics covered in this chapter to develop a simple model for the Earth's climate with which we can begin to understand how humans can alter the climate.

3.5 Chapter summary

- Energy is expressed in units of joules (J). Power is the rate that energy is flowing, and it is express in watts (W); 1 W = 1 J/s.
- Temperature is a measure of internal energy of an object and is frequently expressed by physicists in units of Kelvin. The temperature in degrees Kelvin is equal to the temperature in degrees Celsius plus 273.15.

- Photons are small discrete packets of energy. They have a characteristic size, known as the wavelength, which determines how the photons interact with matter. Photons with wavelengths between 0.3 and 0.8 μm are visible to humans; photons with wavelengths between 0.8 and 1,000 μm are known as infrared.
- Most objects emit blackbody radiation. The characteristic wavelength emitted by a blackbody is equal to $3,000/T$ (where wavelength is in micrometers and temperature is in degrees Kelvin). The total power emitted per unit area by a blackbody is equal to σT^4, where $\sigma = 5.67 \times 10^{-8}$ (W/m^2)/K^4 and temperature is in degrees Kelvin. Photons emitted by room-temperature objects are in the infrared and not visible to humans.
- When a photon is emitted by an object and then absorbed by another object, this process transfers a small amount of energy from the emitter to the absorber.
- If the energy received by an object by absorbing photons exceeds the energy lost by emitting photons, then the object's internal energy increases – and it warms up. The object cools off if the energy in emitted photons exceeds the energy received by absorbing photons.

Additional reading

For additional information about blackbody radiation, consult an introductory physics book. Most have sections on blackbody radiation.

Terms

Blackbody
Electromagnetic radiation
Emissions spectra
Energy
Energy balance
Equilibrium
Infrared radiation
Internal energy
Joule
Micrometer
Photons
Power
Temperature
Ultraviolet
Visible photons
Watt
Wavelength

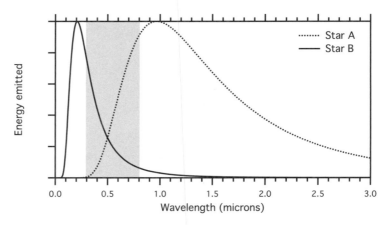

Fig. 3.7   Emissions spectra of two hypothetical stars.

Problems

1. The temperature of an object goes up by 1 K. How much did it go up in degrees Fahrenheit and how much in degrees Celsius?
2. A sphere with a radius of 1 m has a temperature of 100 °C. How much power is it radiating? Remember that temperatures have to be converted to degrees K.
3. As a room-temperature object increases in temperature, it begins to glow. Describe the progression in colors as the object heats up. Ultimately, what happens to the glow if the warming continues to nearly infinite temperatures?
4. Consider two stars that have the spectra shown in Figure 3.7. Based just on the information provided in this plot, what are the colors and radiating temperatures of the stars? (The gray shading shows the range of wavelengths that humans can see.)
5. How much total energy (in watts) is the Sun radiating? It is a 6000-K blackbody with a radius of 700,000 km.
6. You can dim an incandescent bulb by decreasing the temperature of the filament. What do you think happens to the color of the bulb as it dims? Find a dimmer and test your hypothesis.
7. If you run a 60-W light bulb for 1 week, how many joules of energy have been consumed?
8. Why are incandescent light bulbs being phased out in many countries (including the United States)?
9. The Sun as a blackbody:
    a. The Sun is a 6000-K blackbody. At what characteristic wavelength does it radiate?
    b. At what characteristic wavelength does a blackbody at room temperature radiate?
    c. How much power per unit area is the Sun radiating?
10. Note that E_{in} is the energy being absorbed by an object, and E_{out} is the energy being radiated:

a. If the temperature of an object is not changing, what does this tell us about E_{in} and E_{out}?

b. If the temperature of an object is increasing, what does this tell us about E_{in} and E_{out}?

11. Your bank account has the same balance on April 1 as it did on March 1. Your friend suggests that this means that you did not deposit or withdraw any money for the entire month. Is that correct? Explain why or why not.

A simple climate model

Scientists have been studying the Earth's climate for nearly 200 years. Over that time, and especially over the past 30 years, a sophisticated and well-validated theory of our climate has emerged. In this chapter, we take the fundamental physics we learned in Chapter 3 and use it to explain how greenhouse gases warm the planet and why the temperature of the Earth is what it is. By the end of the chapter, we will understand why scientists have such high confidence that adding greenhouse gases to the atmosphere will warm the planet.

4.1 The source of energy for our climate system

The first step to understanding the climate is to do an energy budget calculation: What is the energy in and energy out for the Earth? The ultimate source of energy for our planet is the Sun, which puts out an amazing 3.8×10^{26} W (380 trillion trillion W) of power. The Sun emits photons in all directions, so only a small fraction of the photons emitted end up falling on the Earth. To determine the exact amount of solar energy hitting the Earth, imagine a sphere surrounding the Sun, with a radius equal to the Sun–Earth distance, or 150 million km (Figure 4.1). Because the sphere completely encloses the Sun, all of the sunlight emitted by the Sun must fall on the interior of the sphere. The surface area of the sphere is $4\pi r^2 = 4\pi(150 \text{ million km})^2 = 2.8 \times 10^{17}$ km$^2 = 2.8 \times 10^{23}$ m^2. This means that the energy emitted by the Sun falling on a 1-m^2 surface of the sphere is 3.8×10^{26} W $\div 2.8 \times 10^{23}$ m$^2 = 1{,}360$ W/m^2. This value, 1,360 W/m^2, is known as the *solar constant* for the Earth; it is represented in certain equations by the symbol S.

An example: What is the solar constant for Venus?

Our next-door neighbor Venus is located 108 million km from the Sun. Thus, the solar constant for Venus is 3.8×10^{26} W divided by the surface area of a sphere with a radius of 108 million km, 1.47×10^{23} m^2 – which yields a value of 2,600 W/m^2.

Another way to calculate this is to take the Earth's solar constant and multiply by the squared ratio of the radius of the Earth's orbit to Venus' orbit, $(150/108)^2$. In general, the solar constant scales as $1/r^2$, so that a planet with half the orbital radius has a solar constant that is four times larger whereas a planet with twice the orbital radius has a solar constant one fourth as large.

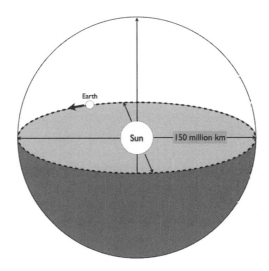

Fig. 4.1   Solar constant calculation: A sphere (gray) surrounds the Sun with a radius equal to the Earth's orbit (dashed line); all radiation emitted by the Sun (black arrows) falls on this sphere.

Now that we know how much energy is falling on every square meter at the Earth's orbit, we can now answer the question of how much solar energy is falling on the Earth. The easiest way to quantitatively calculate this is to realize that, if we set up a screen behind the Earth, the Earth would cast a circular shadow on the screen, and the shadow would have a radius equal to the radius of the Earth (Figure 4.2). The amount of sunlight falling on the Earth is equal to the amount that would have fallen into the shadow area if the Earth were not there. This is equal to the area of the shadow, πR^2, where R is the radius of the Earth, times the solar constant S.

Given that the radius of the Earth is approximately 6,400 km = 6.4×10^6 m, and $S = 1,360$ W/m^2, solar energy is falling on the Earth at a rate of 1.8×10^{17} W or 180,000 TW (1 TW is a terawatt, which is a trillion watts). This is an immense amount of power. Humans today consume 15 TW or so, so this simple calculation shows why solar energy is so attractive: If we could capture just 0.01% of the solar energy falling on the Earth, we could satisfy all of the world's current energy needs.

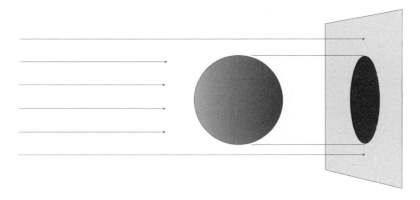

Fig. 4.2   The Earth is casting a shadow on a screen placed right behind it because it blocks sunlight. The total amount of solar energy falling on the Earth is the same as what would have fallen into the shadow area.

Not all of the photons from the Sun that fall on the Earth are absorbed by it. A fraction of the solar energy is reflected back to space by clouds, ice, and other reflective elements of the Earth system. The reflectivity of a planet is called the *albedo*, from the Latin word for "whiteness."[1] In equations, the albedo is usually represented by the Greek letter α (alpha). Thus, the total rate of energy in (E_{in}) for the Earth is

$$E_{in} = S(1 - \alpha)\pi R^2 \qquad (4.1)$$

Here α is the fraction of photons that are reflected, so $1 - \alpha$ is the fraction of photons that are absorbed. For the Earth, $\alpha = 0.3$, meaning that about 126,000 TW are absorbed by the Earth system and the other 54,000 TW of solar photons are reflected back to space. In the rest of the chapter, we will find it more useful to do the calculation per square meter of the Earth's surface area:

$$\frac{E_{in}}{\text{area}} = \frac{S(1 - \alpha)\,\pi R^2}{4\pi R^2} = \frac{S(1 - \alpha)}{4} \qquad (4.2)$$

Note that the πR^2 terms cancel, so that the net amount of solar energy absorbed by the Earth per square meter of the Earth's surface is *not* a function of the Earth's size. Plugging values of S and α into Equation 4.2, we obtain a value of 238 W/m^2 for the Earth.

It is worth pointing out that I have become a bit sloppy with the terms *energy* and *power*. In Equation 4.2, for example, the mathematical abbreviation for *energy in* appears on the left-hand side, but the right-hand side has units of watts, the unit of power. The physics pedants will argue that we should be writing *power in* instead of *energy in*, and they are indeed correct. However, my choice of terminology here reflects the terminology actually used by scientists who do these kinds of energy-balance calculations. If you went to a meeting of climate scientists, or read the peer-reviewed climate literature, you find that they use *power* and *energy* interchangeably in equations like 4.2.

So the Earth absorbs an average of 238 W/m^2 from the Sun, but that does not mean that every square meter of the Earth absorbs this amount. In fact, the amount of solar energy absorbed varies widely across the planet. First, at any given time, half of the planet is experiencing night, during which time that half is receiving no energy from the Sun. During daytime, the amount of sunlight falling on a square meter is determined by the orientation of that square meter with respect to the incoming beams of sunlight. The amount of sunlight received is at maximum if the surface is oriented perpendicular to the incoming beam (Figure 4.3a). As the surface rotates away from perpendicular to the beam of photons, the amount of solar energy intercepting the surface decreases (Figure 4.3b), eventually reaching zero for a surface parallel to the incoming beam (Figure 4.3c).[2]

Figure 4.4 shows how this fact leads to variations in the amount solar energy falling on the Earth's surface with latitude. In the tropics (Arrow A), the surface of the Earth

---

[1] The word *albino* comes from the same Latin word.

[2] For those with a good grasp of geometry, the energy falling on a square meter of surface falls off as the cosine of the latitude (or the angle of the surface from perpendicular).

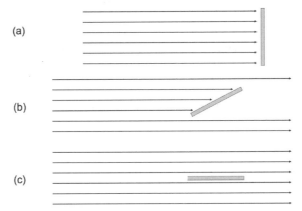

Fig. 4.3 Schematic showing how the amount of energy falling on a surface is dependent on the angle between the surface and the incoming beams of light: (a) perpendicular, (b) rotated away from perpendicular, and (c) parallel.

is perpendicular to the incoming solar light beams, corresponding to the situation in Figure 4.3a. This means that, on average, more solar energy falls on the tropics than on higher latitudes. The surface in the mid-latitudes (Arrow B) is at a moderate angle to the incoming solar light beams, corresponding to the situation in Figure 4.3b. Thus, the mid-latitudes receive less solar radiation per square meter than the tropics. Finally, the polar regions (Arrow C) correspond to the situation in Figure 4.3c, so this region receives even less solar energy.

In addition to variations in the incoming sunlight with latitude, the albedo of the planet also varies widely. The tropics are mainly open ocean, which is dark and therefore has a low albedo. Combined with the large amount of solar energy per square meter, the tropics therefore experience far more solar heating than anywhere else on the planet. This provides us with a simple but fundamentally correct explanation of why the tropics tend to be the warmest place on the planet. The high latitudes are frequently covered by ice, giving them a high albedo. Combined with the small amount of solar energy received per square meter, this means that the polar regions experience the smallest amount of solar heating and therefore tend to be the coldest place on the planet.

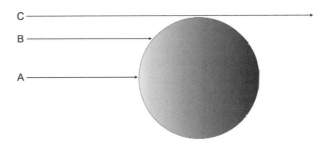

Fig. 4.4 Schematic showing how the amount of solar energy falling on a square meter of the Earth's surface is determined by the latitude.

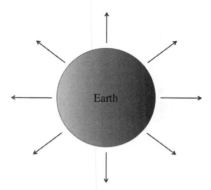

Fig. 4.5   Diagram of the Earth, showing the emission of infrared blackbody radiation in all directions.

4.2 Energy loss to space

In the early 19th century, Joseph Fourier, one of history's great mathematicians, asked a deceptively simple question: Because energy is always falling on the Earth from the Sun, why doesn't the Earth heat up until it is the same temperature as the Sun? The answer he determined is that the Earth is losing energy at a rate equal to the rate at which it is receiving energy from the Sun – otherwise, the Earth would indeed be continuously heating up.

On the basis of what we learned in Chapter 3, you may rightly guess that the Earth loses energy back to space by means of blackbody radiation (Figure 4.5). For a blackbody, P/a is σT^4, where P/a is the power emitted per square meter, T is the temperature of the planet, and σ is the Stefan–Boltzmann constant, 5.67×10^{-8} $(W/m^2)/K^4$. Setting the rate of energy in (Equation 4.2) equal to the rate of energy out, we get the following equation:

$$\frac{S(1-\alpha)}{4} = \sigma T^4 \tag{4.3a}$$

Solving for T, we get

$$T = \sqrt[4]{\frac{S(1-\alpha)}{4\sigma}} \tag{4.3b}$$

Plugging $S = 1{,}360$ W/m^2 and $\alpha = 0.3$ into Equation 4.3b yields[3] a temperature T of 255 K ($-18\,°$C). The actual average temperature of the Earth is closer to 288 K (15 °C), so our estimate of the Earth's temperature is too cold. So where did our calculation go wrong? It turns out that what we have neglected is the heating of the planet by the Earth's atmosphere, which is frequently referred to as the *greenhouse effect*.

---

[3] The mathematical equation $a = \sqrt[4]{y}$ means that $a^4 = y$. To calculate the fourth root of y on a calculator, you can use the y^x key found on most calculators, where $x = 0.25$. A simpler way to calculate the fourth root of y is to take the square root of y, and then take the square root of that number – in other words, $a = \sqrt{\sqrt{y}}$.

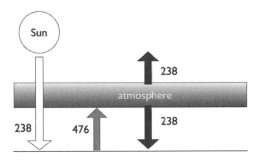

Fig. 4.6   Schematic of energy flow on a planet with a one-layer atmosphere. The atmosphere is represented by a single layer that is transparent to visible photons but absorbs all infrared photons that fall on it. The arrows show global average energy flows with values in W/m^2.

4.3 The greenhouse effect

4.3.1 One-layer model

To understand the impact of the atmosphere on our planet's temperature, let's make the following assumptions (which turn out to be reasonably accurate). The Earth's atmosphere is transparent to visible photons emitted by the Sun (which have wavelengths from 0.3–0.8 μm), so these photons speed through the atmosphere and reach the surface. The atmosphere is opaque to infrared photons emitted by the surface (wavelengths from 4–20 μm), and so all of these photons are absorbed by the atmosphere. The atmosphere also behaves like a blackbody, so it emits photons based on its temperature. Half of these emitted photons are emitted in the upward direction, toward space. These photons escape to space and carry energy away from the Earth. The other half are emitted downward, and these photons are all absorbed by the surface. This situation is diagrammed in Figure 4.6. For conceptual simplicity, the diagram shows the effects of the atmosphere concentrated in a single thin layer, which is why this model is frequently called a "one-layer" model.

To calculate the surface temperature in this model, we first assume that the temperatures on the planet are not changing, so that the planet as a whole, as well as the surface and the atmosphere individually, must all be in energy balance (where energy in equals energy out). We also assume that the solar constant and albedo for this planet have the same values as for the Earth. Let's first consider the planet as a whole. Energy in to the planet is coming entirely from the Sun. Energy out to space is coming entirely from the atmosphere. Thus, if the energy in from the Sun is 238 W/m^2 (Equation 4.2), then the atmosphere must be radiating 238 W/m^2 upward to space in order for the planet as a whole to be in energy balance. Furthermore, because the atmosphere radiates equally both upward and downward, this means that the atmosphere is also radiating 238 W/m^2 back toward the Earth's surface.

Now let's consider energy balance for the surface. Energy in for the surface is 238 W/m^2 from the Sun and 238 W/m^2 from the atmosphere, for a total of 476 W/m^2. This means that the surface has to be emitting 476 W/m^2 upward in order to achieve energy balance.

Finally, let's double-check the energy balance for the atmosphere. Energy in comes from the surface, which is emitting 476 W/m^2. Energy out comes from emission of 238 W/m^2 upward to space and 238 W/m^2 downward to the surface, for a total energy out of 476 W/m^2. Thus, the atmosphere is indeed in energy balance.

So what is the temperature of the surface? If we know the energy out for the surface, we can determine its temperature by using the Stefan–Boltzmann equation from Chapter 3: Energy out or $E_{out} = P/a = \sigma T^4$. Solving $\sigma T^4 = 476$ W/m^2 for T yields a surface temperature of 303 K (30 °C), which is 48 °C warmer than that for the planet without an atmosphere. Note that we have assumed that a single temperature characterizes the entire surface, which is equivalent to assuming that energy is transported very rapidly around the planet.

Thus, the addition of an atmosphere that is opaque to infrared radiation has significantly warmed the planet's surface. Conceptually, this occurs because the surface of the planet with an atmosphere is heated not just by the Sun but also by the atmosphere. Of course, if you walk outside, you cannot see the atmosphere heating the Earth's surface because the photons it emits are not visible, but they still carry energy. When scientists talk about the greenhouse effect, it is this heating of the surface by the atmosphere to which they are referring.

An alternative way to think about the greenhouse effect is that the atmosphere warms the surface by making it harder for the surface to lose energy to space. Without an atmosphere, all of the photons emitted by the surface escape to space; the surface has to emit only 238 W/m^2 for the planet to be in energy balance. With a one-layer atmosphere, though, only half of the energy emitted by the surface ends up escaping to space – the other half is returned to the surface. This means that the surface must emit twice as much, 476 W/m^2, in order for 238 W/m^2 to escape to space. This in turn means that a planet with an atmosphere must have a warmer surface than a planet without an atmosphere.

4.3.2 Two-layer model

Now let's consider a planet with a thicker atmosphere. In this case, photons emitted by the surface are absorbed in the lower atmosphere. Photons emitted by the lower atmosphere in the upward direction are absorbed by the upper atmosphere. If they are emitted by the upper atmosphere in the upward direction, then they escape to space. Photons emitted in the downward direction by the lower atmosphere are absorbed by the surface, while photons emitted in the downward direction by the upper atmosphere are absorbed by the lower atmosphere. This is known as a two-layer model; it is shown schematically in Figure 4.7.

Once again, the key to determining the surface temperature is to enforce energy balance for the planet as a whole, the surface, and both atmospheric layers. The easiest way to do this is to work downward from the topmost layer to the surface. We know that planetary energy balance requires energy out for the planet to balance energy in from the Sun. In addition, because energy out comes entirely from the upper layer, that means that it must be emitting 238 W/m^2 to space in order to balance the 238 W/m^2 that the Sun is providing the planet. That, in turn, means that the upper layer is also emitting 238 W/m^2 downward.

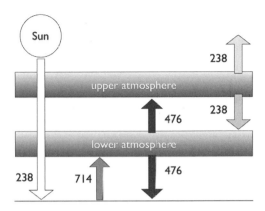

Fig. 4.7 Schematic of energy flow on a planet with a two-layer atmosphere, with values in W/m^2.

Totaling the emissions in both directions, we find that the upper layer is emitting 476 W/m^2. Because energy out must equal energy in for the layer, this layer must be receiving 476 W/m^2 from the lower atmospheric layer. Thus, we know the lower layer is emitting 476 W/m^2 upward – and therefore downward, too, for a total of 952 W/m^2. For the lower layer to achieve energy balance, the lower layer must also be receiving 952 W/m^2. We already calculated that 238 W/m^2 are coming from the upper layer, so that means that 714 W/m^2 must be coming from the surface to the lower layer.

Finally, for the surface to be emitting 714 W/m^2, the surface temperature must be 335 K (62 °C), which is 32 °C warmer than the planet with a thinner one-layer atmosphere. Thus, by adding a second layer to the atmosphere and making it thicker, we have further increased the surface temperature of the planet.

4.3.3 *n*-layer model

Now let's derive the surface temperature for a planet with n layers (Figure 4.8). For some variety, let's assume that the planet has a solar constant of $S = 2,000$ and an

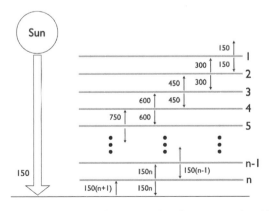

Fig. 4.8 Schematic of energy flow on a planet with an *n*-layer atmosphere; layers are numbered from 1 to *n* (topmost to bottommost layers), with values in W/m^2.

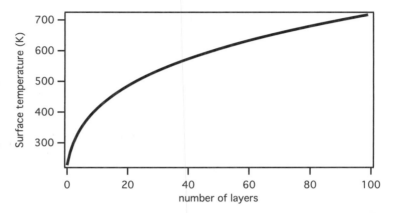

Fig. 4.9 Surface temperature for the n-layer planet, as a function of the number of layers. Here $S(1 - \alpha)/4 = 150$ W/m^2.

albedo of $\alpha = 0.7$. Thus, energy in for this planet is $S(1 - \alpha)/4 = 150$ W/m^2. This means that upward emissions from the topmost layer of the atmosphere (Layer 1) must also be 150 W/m^2. Moreover, because upward and downward emissions must be the same, this layer is also emitting 150 W/m^2 downward – so that total energy out for this layer is 300 W/m^2. This in turn means that energy in for Layer 1 must also be 300 W/m^2.

The energy in for Layer 1 comes entirely from upward emissions of Layer 2. Layer 2 must therefore be emitting 300 W/m^2 upward. This means that it is also emitting 300 W/m^2 downward, for a total energy out of 600 W/m^2. Energy in for Layer 2 comes from downward emissions of Layer 1 and upward emissions of Layer 3 and must total 600 W/m^2 in order to balance energy out. Downward emissions from Layer 1 are 150 W/m^2, which means that upward emissions from Layer 3 must be 450 W/m^2.

Layer 3 must be emitting 450 W/m^2 both upward and downward, for a total energy out of 900 W/m^2. Energy in from downward emissions from Layer 2 is 300 W/m^2, meaning that upward emissions from Layer 4 must be 600 W/m^2.

By this time, you may see a pattern emerging, and we can extrapolate to the bottommost layer, layer n. For layer n, both upward and downward emissions are 150n. This in turn means that the surface is receiving 150n emitted from the bottommost layer and 150 W/m^2 from the Sun. For energy to balance, the surface must be emitting $150(n + 1)$ W/m^2 upward. Setting $150(n + 1) = \sigma T^4$, we can solve for the surface temperature T of this planet:

$$T = \sqrt[4]{\frac{150\,(n + 1)}{\sigma}} \tag{4.4}$$

Figure 4.9 shows the surface temperature as a function of n, calculated by using Equation 4.4. As you increase the number of layers, the atmosphere gets hotter and hotter. However, the warming due to one additional layer decreases as the amount of greenhouse gas in the atmosphere increases.

Planet	Solar constant (W/m²)	Albedo	Observed surface temperature (K)	Inferred n
Table 4.1 Data on the four inner planets in our solar system			
Mercury	10,000	0.1	452	0.052
Venus	2,650	0.7	735	82
Earth	1,360	0.3	289	0.65
Mars	580	0.15	227	0.22

We can also write the general solution for the surface temperature of an n-layer planet:

$$T = \sqrt[4]{\frac{(n+1)\,S\,(1-\alpha)}{4\sigma}} \tag{4.5}$$

This is an important equation and we will return to it repeatedly throughout the rest of this book. It says that the surface temperature of the planet is basically determined by three parameters: the number of layers in the atmosphere (n), the solar constant (S), and the albedo (α). To connect this equation to the real world, I should make clear what the "number of layers" physically represents. As I will discuss in Chapter 5, it is the *greenhouse gases* in our atmosphere that absorb infrared photons. And the number of layers is equivalent to the amount of greenhouse gas in the atmosphere. Therefore, an increase in the greenhouse gas in the atmosphere corresponds to an increase in the number of layers – and a warming climate.

4.4 Testing our theory with other planets

It is important for me to emphasize that the n-layer model discussed in Subsection 4.3.3 ignores several important physical processes. For example, not all energy transport on a planet is by radiation (some is transported by atmospheric motions, such as deep convective thunderstorms); the model assumes an infinitely fast horizontal energy transport, which is not correct; and the model assumes that the atmosphere absorbs all infrared photons, which is not correct. This means that you should not expect the model to produce quantitatively accurate surface temperatures.

Nonetheless, as I will show in this section, the model is successful in making qualitative projections of planetary temperatures. Here we will test whether Equation 4.5 explains the relative surface temperatures of the Earth and its nearby neighbors, namely Mercury, Venus, and Mars. Table 4.1 lists the important characteristics of the planets, and it reveals some puzzles. Mercury is the planet closest to the Sun, yet Venus, twice as far from the Sun as Mercury, has a surface temperature that is approximately 300 K warmer. This result becomes even more puzzling when we realizes that, because of its high albedo, the energy in for Venus, $S(1-\alpha)/4 = 200$ W/m², is more than a factor of 10 smaller than that for Mercury (2,250 W/m²). It is even less than the energy in for the Earth (238 W/m^2) – yet Venus is approximately 450 K hotter than the Earth.

Given the surface temperature, albedo, and solar constant, we can solve Equation 4.5 for n, which is the number of layers each planet requires to satisfy energy balance. This "inferred n" is also listed in Table 4.1. The inferred n for Mercury is near zero, suggesting it has almost no greenhouse effect. This is correct, because Mercury has essentially no atmosphere. For Mars, inferred $n = 0.22$. This again makes some sense – Mars has a thin atmosphere containing carbon dioxide, so it does have some greenhouse effect. However, the Martian atmosphere has fewer greenhouse gases than the Earth's atmosphere, so the greenhouse effect on Mars is expected to be weaker than that on Earth. Our calculations confirm that.

Finally, our calculations suggest that Venus, with inferred $n = 82$, has a massive greenhouse effect. This is again correct. The surface pressure on Venus is 90 times that of Earth (1,300 psi, or pounds per square inch, compared with 14.5 psi here on Earth), and the atmosphere is mainly composed of carbon dioxide. The result of this massive, greenhouse-gas-rich atmosphere is a planet hotter than the inside of your oven on broil – hot enough even to melt lead. Thus, we conclude that Equation 4.5 has successfully explained the relative temperatures of the innermost planets of our solar system.

4.5 Chapter summary

- In this chapter, we created a very simple climate model based on the fact that the solar energy received by a planet (E_{in}) must be balanced by the energy that is radiating to space (E_{out}).
- For a planet, $E_{in} = S(1 - \alpha)/4$. S is the solar constant, which is the amount of energy per square meter that sunlight is delivering at the planet's orbit, and α is the planet's albedo, which is the fraction of photons that fall on the planet that are reflected back to space.
- The energy out for a planet is due to blackbody radiation; $E_{out} = \sigma T^4$.
- In our simple model of the climate, the atmosphere is entirely transparent to visible radiation from the Sun, but it absorbs all infrared radiation. We can then calculate the surface temperature by enforcing energy balance for the surface, the atmosphere, and the planet as a whole.
- We derived a general equation, Equation 4.5, for the surface temperature T of a planet. It is repeated here:

$$T = \sqrt[4]{\frac{(n + 1)\, S\, (1 - \alpha)}{4\sigma}}$$

- This equation says that the surface temperature of the planet is basically determined by three parameters: the number of layers in the atmosphere (n), which is a proxy for how much greenhouse gas is in the atmosphere; the solar constant (S); and the albedo (α).

• This simple model also explains the relative temperatures of the Earth's nearest neighbors, namely Mercury, Venus, and Mars.

Additional reading

For a more complete description of climate physics, see the following books.

J. T. Houghton, *The Physics of Atmospheres* (Cambridge: Cambridge University Press, 2001).
D. L. Hartmann, *Global Physical Climatology* (New York: Academic Press, 1994).

Terms

Albedo
Greenhouse effect
Greenhouse gases
Solar constant

Problems

1. What is the surface area of a sphere with radius r? What is the area of a disk with radius r?
2. A planet in another solar system has a solar constant of $S = 2,000$ W/m^2, and the distance between the planet and the star is 100 million km.
   a) What is the total power output by the star? (Give your answer in watts.)
   b) What is the solar constant of a planet located 75 million km from the same star? (Give your answer in watts per square meter.)
3. Draw a diagram (like Figure 4.6) that shows the energy flows for a planet with a one-layer atmosphere. The solar constant for the planet is $S = 900$ W/m^2 and the albedo of the planet is $\alpha = 0.25$. Make sure each arrow is labeled with the energy flow. What is the surface temperature of this planet?
4. Draw a diagram (like Figure 4.7) that shows the energy flows for a planet with a two-layer atmosphere. The solar constant for the planet is $S = 3,000$ W/m^2 and the albedo of the planet is $\alpha = 0.1$. Make sure each arrow is labeled with the energy flow. What is the surface temperature of this planet?
5. Two people argue about why Venus is so much warmer than the Earth. The first argues that it's because Venus is closer to the Sun, so it absorbs more solar energy. The second argues that it's because Venus has a thick, greenhouse-gas-rich atmosphere. Which person is right, and why is the other one wrong?

6. Some recently discovered planets in other solar systems are so hot that they glow in the visible; they are literally "red hot" (e.g., do a Google search for "HD 149026b").

   a) How many atmospheric layers would the Earth need before it glowed in the visible? (Assume $S = 1,360$ W/m^2 and $\alpha = 0.3$.) To answer this, you must first estimate what temperature the Earth has to be to begin glowing.

   b) Alternatively, what would the solar constant have to increase to for a one-layer planet with an albedo of $\alpha = 0.3$?

   c) How far would the Earth have to be from the Sun in order to have this solar constant?

7. Assume a planet with a one-layer atmosphere has a solar constant of $S = 2,000$ W/m^2 and an albedo of $\alpha = 0.4$.

   a) What is the planet's surface temperature? Make the standard assumption that the atmosphere is transparent to visible photons but opaque to infrared photons.

   b) During a war on this planet, a large number of nuclear weapons are exploded, which kicks enormous amounts of dust and smoke into the atmosphere. The net result is that the atmosphere now absorbs visible radiation – so solar energy is now absorbed in the atmosphere. Draw a diagram like Figure 4.6 to show the fluxes for this new situation, and calculate the planet's surface temperature. The solar constant and albedo remain unchanged.

   c) Explain in words why the temperature changes the way it does after the nuclear war. Is describing this as "nuclear winter" appropriate?

8. Assume a planet with a one-layer atmosphere and values of solar constant of $S = 1,000$ W/m^2 and albedo of $\alpha = 0.25$. Let's assume there's some dust in the atmosphere, so that 50% of the Sun's energy is absorbed by the atmosphere and 50% by the surface. Draw a diagram like Figure 4.6 to show the fluxes, and calculate the planet's surface temperature.

9. Derive an expression for the fraction of energy received by the surface that comes from the atmosphere (this is the amount of energy that comes from the atmosphere divided by the sum of energy from the Sun and energy from the atmosphere). Using values in Table 4.1, calculate the fraction for Mercury, Earth, and Venus. Make the standard assumption that the atmosphere is transparent to visible photons but opaque to infrared photons.

10. On Mercury, which has no atmosphere, the difference in temperature between daytime and nighttime temperatures can be 700 K. On the Earth, the difference between daytime and nighttime temperatures can be 30 K. On Venus, there is basically no difference between daytime and nighttime temperatures. Why is this? (If you get stuck, working Question 9 might help you answer this question.)

11. As we will discover in Chapter 11, one way to solve global warming is to increase the reflectivity of the planet (I will explain how later). To reduce the Earth's temperature by 1 K, how much would we have to change the albedo? (assume a one-layer planet with an initial albedo of 0.3 and solar constant of 1360 W/m^2).

12. Given fixed n and α, how does the temperature of a planet vary with r, the distance between the planet and the star? Hint: Work out how S varies with r, and plug that into Equation 4.5.

13. A planet has a solar constant of $S = 2{,}000$ W/m^2, an albedo of $\alpha = 0.7$, and a radius of $r = 3{,}000$ km. What would happen to the temperature if the planet's radius doubles?

14. One argument you hear against mainstream climate science is that adding greenhouse gases to the atmosphere is like painting a window. Eventually, the window is opaque so that adding another coat of paint does nothing. Is this a good analogy? Is there a point where adding greenhouse gases to the atmosphere does not lead to increases in the planet's temperature?

5 The carbon cycle

In the simple model of the climate derived in Chapter 4, the temperature of a planet is set by the number of atmospheric "layers," the albedo, and the solar constant. I said there that the number of layers was determined by the abundance of greenhouse gases in the atmosphere, but I did not define what a greenhouse gas is, or which components of our atmosphere are greenhouse gases. In this chapter, I address these questions and discuss in detail one of our atmosphere's most important greenhouse gases, namely carbon dioxide.

Carbon dioxide, or CO_2, is the primary greenhouse gas emitted by human activities, and policies to control modern climate change frequently focus on reducing our emissions of this gas. Understanding carbon dioxide requires more than just knowing how much of it humans are dumping into the atmosphere. It also requires an understanding of the *carbon cycle* – how carbon moves between the atmosphere, ocean, land biosphere, and rocks on the Earth. This will help us understand what happens to carbon dioxide after it is emitted into the atmosphere, which in turn will help us understand the future trajectory of our climate.

5.1 Greenhouse gases and our atmosphere's composition

As we learned in Chapter 4, the greenhouse effect occurs because our atmosphere is basically transparent to visible photons but absorbs infrared photons. It turns out that only a few of the components of our atmosphere actually absorb infrared photons, and it is these *greenhouse gases* that are responsible for the Earth's greenhouse effect. In this section, I describe the composition of our atmosphere, with a particular focus on greenhouse gases.

Approximately 78% of the dry atmosphere ("dry atmosphere" does not include water vapor)[1] is made up of diatomic nitrogen or N_2 – two nitrogen atoms bound together. Roughly 21% is diatomic oxygen or O_2, which is two oxygen atoms bound together. Oxygen is the part of the atmosphere that we need to breathe to survive. Argon atoms make up approximately 1% of our atmosphere. None of these three constituents, which together make up more than 99% of the dry atmosphere, absorbs infrared photons, so they are not greenhouse gases; therefore, they do not warm the surface of the planet.

---

[1] Throughout this section, the percentages given are of volume, not mass. For the chemists reading this, this is the same as mole fraction.

The next biggest component of the atmosphere is water vapor or H_2O, a constituent whose abundance varies widely from place to place. In the warm tropics, water vapor can make up as much as 4% of the atmosphere. In cold polar regions, in contrast, water vapor may make up only 0.2%. Its abundance decreases rapidly with altitude, and in the stratosphere it typically makes up 0.0005% of the atmosphere.

Water vapor is the most abundant and important greenhouse gas in our atmosphere. Its main source is evaporation from the oceans, and it is primarily removed from the atmosphere when water forms raindrops and these fall to the surface. Human emissions of water vapor contribute essentially nothing to its atmospheric abundance. In Chapter 6, I will talk in more detail about the role water vapor plays in climate change and how humans are indirectly increasing its abundance.

Taken together, diatomic nitrogen and oxygen, water vapor, and argon make up more than 99.95% of the atmosphere. You might expect the remaining 0.05% to have no important role because it seems like such a small amount – but you would be wrong. This last smidgen of atmosphere is crucial to life on the planet.

The largest part of this remaining 0.05% is carbon dioxide or CO_2, which made up 0.039% of the atmosphere in 2010. Carbon dioxide absorbs infrared photons and is therefore a greenhouse gas. In fact, it is the second most important greenhouse gas, behind water vapor. Note that 0.039% is an awkwardly small number, so scientists typically express the concentration of trace gases like carbon dioxide in a more convenient unit: *parts per million*. A concentration expressed in parts per million indicates how many molecules out of every million are the gas in question.[2] In this case, 0.039% corresponds to 390 parts per million or ppm, meaning that there are 390 molecules of carbon dioxide in every million molecules of air.

Parts per million can be usefully contrasted to percent, which indicates how many molecules of every 100 are the species in question. In fact, the word *percent* comes from the marriage of the words *per cent*, literally meaning "out of 100." Thus, air is approximately 78% nitrogen, which means that 78 out of every 100 molecules of air are molecules of diatomic nitrogen. Percent is, of course, a widely used concept for expressing fractions, so you frequently hear statements such as "14% of the population is left handed," meaning that 14 out of every 100 people are left handed.

The next most important greenhouse gas in our atmosphere is methane or CH_4. In 2010, it had an atmospheric abundance of 1.8 ppm. Despite its small abundance, methane is also a key player in our climate; I will discuss it in detail in Section 5.6.

Another important greenhouse gas in our atmosphere is nitrous oxide or N_2O, which is present in today's atmosphere at concentrations of about 0.3 ppm. You might know this molecule as "laughing gas," which your dentist might give you before she works on your teeth. It is emitted into the atmosphere from nitrogen-based fertilizer and industrial processes as well as several natural sources.

Halocarbons include chlorofluorocarbons and hydrochlorofluorocarbons, which are synthetic industrial chemicals used as refrigerants (e.g., in air conditioners and refrigerators) and in various other industrial applications. This category also includes

[2] Parts per million can be by volume (number of molecules out of every million) or by mass (grams of constituent out of every million grams of air). Following the previous discussion, all mixing ratios in this book will be by volume.

natural chemicals such as methyl chloride. They are present in today's atmosphere at a concentration of a few parts per billion,[3] and all of them are powerful greenhouse gases.

A final greenhouse gas in our atmosphere is ozone, which is a molecule made up of three oxygen atoms, so its chemical formula is O_3. The abundance of ozone varies widely across the atmosphere – in unpolluted air near the surface, its abundance is about 10 ppb or parts per billion, whereas its abundance can reach 10 ppm in the stratosphere.

Ozone is absolutely necessary for life on our planet because it absorbs high-energy ultraviolet photons emitted by the Sun before they reach the Earth's surface. These photons carry enough energy that they can seriously damage living tissue – leading to diseases such as cancer in humans. Ozone is also one of the primary components of photochemical smog, and breathing it can lead to health problems in humans and animals; ground-level ozone can also damage plants. Thus, you want ozone between yourself and the Sun, but you don't want to breathe it. Because of this, ozone high up in the stratosphere is considered "good" ozone, whereas ozone near the ground is considered "bad" ozone.

In Chapter 6 I will quantify the contribution of these greenhouse gases to climate change. I will just note here that greenhouse gases are not equal in their ability to warm the planet. As a greenhouse gas, methane is roughly 20 times more powerful than carbon dioxide on a per molecule basis – meaning that it takes 20 molecules or so of carbon dioxide to equal the warming from one molecule of methane. The most powerful greenhouse gases on a per molecule basis are the halocarbons. It takes several thousand carbon dioxide molecules to equal the warming from one halocarbon molecule. Don't forget, too, that the three most abundant gases in our atmosphere, namely diatomic nitrogen, diatomic oxygen, and argon, which together make up about 99.9% of the dry atmosphere, are not greenhouse gases at all. These differences in warming potential among various gases have important implications when designing policies to address climate change.

Because of the importance of carbon dioxide to the problem of modern climate change, in the rest of the chapter we will mainly focus on it and the processes that regulate its atmospheric abundance, which are collectively known as the *carbon cycle*.

5.2 Atmosphere–land biosphere–ocean carbon exchange

5.2.1 Atmosphere–land biosphere exchange

We have been directly monitoring the abundance of carbon dioxide in the atmosphere since the middle of the 20th century. Figure 5.1 plots 2 years (24 months) of measurements, which show a distinct seasonal cycle in the abundance of carbon dioxide. During the year, carbon dioxide varies by 6 ppm or so, reaching a maximum in May and a minimum in September.

[3] This means that, of every billion molecules in a volume of air, a few are halocarbons.

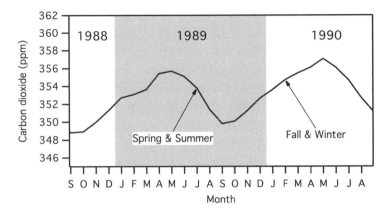

Fig. 5.1   The atmospheric abundance of carbon dioxide from fall 1988 through fall 1990 (data measured at Mauna Loa, Hawaii, and obtained from the NOAA Earth System Research Laboratory/Global Monitoring Division; see ftp://ftp.cmdl.noaa.gov/ccg/co2/trends/co2_mm_mlo.txt).

This annual cycle reflects the annual cycle of plant growth and decay. Plants absorb carbon dioxide from the atmosphere and use it to produce more plant material in a process known as *photosynthesis*:

$$CO_2 + H_2O + \text{sunlight} \rightarrow CH_2O + O_2 \tag{5.1}$$

In this reaction, carbon dioxide, water, and sunlight combine to produce CH_2O and O_2. Note that CH_2O is a combination of carbon and water, and molecules made up of this unit are generally referred to as carbohydrates – in this context, you can think of CH_2O as the chemical formula for a plant. Diatomic oxygen produced in this reaction is released into the atmosphere. This is the main source for the oxygen in our atmosphere, which air-breathing animals like us need to survive.

At the same time, humans, animals, and bacteria consume plant material in order to produce energy through a reaction known as *respiration*:

$$CH_2O + O_2 \rightarrow CO_2 + H_2O + \text{energy} \tag{5.2}$$

The net result of Equation 5.2 is carbon dioxide, which is released back into the atmosphere, and energy, which is used to power the organism. I should emphasize that Equations 5.1 and 5.2 do not represent actual chemical reactions; rather, they represent the net of a large number of complex individual reactions.

It should be apparent that Equation 5.2 is the reverse of Equation 5.1. The carbon dioxide consumed in the production of the plant material in Equation 5.1 is released back into the atmosphere when the plant is consumed in Equation 5.2. Similarly, the oxygen molecule produced in Equation 5.1 is consumed in Equation 5.2. The production of a carbohydrate through photosynthesis followed by its consumption during respiration therefore produces no net change in either carbon dioxide or molecular oxygen. Instead, the net effect is the conversion of sunlight into energy that powers living creatures.

The atmosphere contains approximately 740 gigatonnes of carbon (GtC). A gigatonne is 1 billion metric tons, where 1 metric ton is 1,000 kg or 2,200 lbs. Note that this is just the mass of the carbon atoms in the atmosphere – although the carbon dioxide molecule also contains two oxygen atoms, their mass is not included. Unfortunately, you will sometimes see the mass expressed as the mass of carbon dioxide, which does include the mass of the two oxygen atoms. In that case, the atmosphere contains roughly 2,700 $GtCO_2$ (billion metric tons of carbon dioxide). You can convert between these units by using the fact that 1 GtC = 3.67 $GtCO_2$. In this book, I'll use GtC exclusively, but you must be very careful to recognize whether the mass is given in GtC or $GtCO_2$ when you read anything about climate change.

The land biosphere contains 2,000 GtC or so, stored in living plants and animals, and in organic carbon in soils (e.g., decaying leaves). During a given year, photosynthesis removes approximately 100 GtC from the atmosphere. Respiration closely balances this, transferring 100 billion tons of carbon back to the atmosphere. Thus, over a year, there is no net change in carbon dioxide in the atmosphere or land biosphere as a result of photosynthesis or respiration.

The fact that photosynthesis and respiration are balanced over the year does not mean that they are in balance at every point in time. Most of the Earth's land area – and, therefore, most of the Earth's plants – are found in the northern hemisphere. During the northern hemisphere's spring and summer (May–September), when plants are growing and trees are leafing, global photosynthesis exceeds respiration and there is a net drawdown of carbon dioxide out of the atmosphere and into the land biosphere; we can see this in Figure 5.1.

During the northern hemisphere's fall and winter (October–April), plant material that was produced during the spring and summer decays, releasing carbon back into the atmosphere in the form of carbon dioxide. This means that global respiration exceeds photosynthesis and there is a net transfer of carbon from the biosphere into the atmosphere (which we can also see in Figure 5.1).

An aside: Where does the oxygen in our atmosphere come from?

As I mentioned earlier, photosynthesis followed by respiration is not a net producer or consumer of carbon dioxide or oxygen. Where, then, does the large amount of molecular oxygen in our atmosphere come from? It turns out that it is the result of photosynthesis that is not balanced by respiration. That occurs when a plant grows through photosynthesis, but the plant material is buried before it can be consumed via respiration. When that happens, the oxygen produced during photosynthesis is not consumed. Over the billions of years that life has existed on the planet, this process has built up and now maintains the oxygen levels in our atmosphere.

5.2.2 Atmosphere–ocean carbon exchange

One of carbon dioxide's most important properties is that it readily dissolves in water. Once it has dissolved in water, carbon dioxide can be converted to *carbonic acid* (H_2CO_3) by means of this reaction:

$$CO_2 + H_2O \rightarrow H_2CO_3 \tag{5.3}$$

The carbonic acid can then be converted into other forms of carbon. Because of the conversion of carbon dioxide to many other forms of carbon, the ocean can absorb an enormous amount of carbon dioxide from the atmosphere. Carbon is returned to the atmosphere in a reaction that is the reverse of Equation 5.3:

$$H_2CO_3 \rightarrow CO_2 + H_2O \qquad (5.4)$$

This is followed by the escape of carbon dioxide back into the atmosphere.

This is similar chemistry to that used to make fizzy soft drinks. The manufacturer dissolves large amounts of carbon dioxide into water, thereby producing carbonated water. The conversion in liquid of carbon dioxide to carbonic acid (Equation 5.3) explains why soft drinks tend to be highly acidic. Other flavors and ingredients are also added, and the can or bottle is then sealed to prevent the carbon dioxide from escaping out of the liquid and back into the atmosphere. When the soft drink is opened, the carbon dioxide begins to rapidly escape from the liquid, and in the process forms the little bubbles that give these drinks their characteristic fizziness. By analogy with soda, we can therefore expect that, as the oceans absorb carbon dioxide, they will become more acidic, a topic that we will return to in Chapter 9.

As a result of this chemistry, carbon cycles easily between the atmosphere and ocean. To fully understand this exchange, however, we must think of the ocean as being split into two parts. The first part is the top 100 m or so of the ocean, which exchanges carbon very rapidly with the atmosphere. This part of the ocean is sometimes referred to as the *mixed layer* because it is well mixed by winds and strong weather events, such as hurricanes. This layer contains approximately 1,000 GtC. Below this lies the vast majority of the ocean, and this *deep ocean* exchanges carbon with the mixed layer. The deep ocean also contains most of the ocean's carbon, approximately 38,000 GtC, or 50 times or so more carbon than is in the atmosphere.

5.2.3 The combined atmosphere–land biosphere–ocean system

Figure 5.2 shows a schematic of the combined atmosphere–land biosphere–ocean system. Approximately 100 GtC per year are continuously cycling between the atmosphere and land biosphere as plants absorb carbon dioxide as they grow and then release carbon dioxide when they die. Similarly, approximately 100 GtC per year of carbon atoms are continuously dissolving into the ocean's mixed layer, while the same mass of carbon atoms is coming out of the ocean and back into the atmosphere, thereby cycling between the atmosphere and ocean.

In addition, the mixed layer is exchanging 100 GtC or so with the deep ocean. This occurs as ocean currents transport water carrying carbon from the mixed layer to and from the deep ocean. It also occurs when sinking organic matter, such as dead organisms or fecal material, travels from the mixed layer into the deep ocean. This is sometimes referred to as the biological carbon pump.

To get an idea of how fast this exchange occurs, we can calculate *turnover times* for the atmosphere and land biosphere reservoirs. The turnover time for the atmosphere is the length of time that a carbon atom in the atmosphere will remain there before being transferred into one of the other two reservoirs. We can roughly estimate this

The carbon cycle

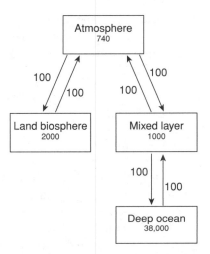

Fig. 5.2 A schematic of exchange between the atmosphere, land biosphere, and ocean. Reservoirs are in GtC; fluxes are in GtC/yr.

as the size of the reservoir, 740 GtC, divided by the total flux out of the reservoir, 200 GtC/yr (100 GtC/yr goes into the land biosphere and 100 GtC/yr goes into the mixed layer). This yields an atmospheric turnover time of 3.7 years. This turnover time is also referred to as a "lifetime" or "residence time."

This means that a carbon atom will stay in the atmosphere for only 4 years or so before it is transferred into the land biosphere or ocean. Remember that this is an average value – an individual molecule of carbon dioxide may remain in the atmosphere for a shorter or longer time. Another way to think about a turnover time is that, over a period of 4 years, enough exchange will have taken place to replace all[4] of the carbon that is in the atmosphere with carbon from the land biosphere or ocean.

The turnover time of carbon in the land biosphere is 2,000 GtC ÷ 100 GtC/yr = 20 years. This means that a carbon atom in the land biosphere will stay there for approximately 20 years before being transferred into the atmosphere. Thus, it takes a few decades for a carbon atom to make a round trip from the land biosphere into the atmosphere and back into the land biosphere.

We can also calculate the turnover times for the ocean reservoirs. The total flux out of the mixed layer is 200 GtC/yr (100 GtC/yr is exchanged with the atmosphere and 100 GtC/yr is exchanged with the deep ocean), so the turnover time for the mixed layer is 1,000 GtC ÷ 200 GtC/yr = 5 years. The turnover time for the deep ocean is several centuries; 38,000 GtC ÷ 100 GtC/yr = 380 years. Thus, it takes several centuries for a carbon atom to make a round trip from the atmosphere through the mixed layer to the deep ocean, and back.

Another way to think about this is that the atmosphere exchanges carbon rapidly (*time scale* of years) with the land biosphere and mixed layer, and much more slowly

---

[4] This turnover time is actually an *e*-folding time. After one *e*-folding time, $1/e = 37\%$ of the original carbon remains. However, we can roughly think that, after one *e*-folding time, all of the carbon has been replaced.

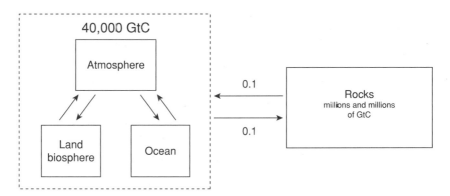

Fig. 5.3 A schematic of exchange between the atmosphere–land biosphere–ocean reservoir and the rock reservoir (reservoirs are in GtC; fluxes are in GtC/yr).

(time scale of centuries) with the deep ocean. In Chapter 8, I will explain why this is so important for the climate change problem.

5.3 Atmosphere–rock exchange

Most of the carbon in the world – that is, many millions of gigatonnes of carbon – is stored in rocks, such as limestone ($CaCO_3$). This carbon is slowly exchanging with the atmosphere–land biosphere–ocean system (Figure 5.3). Carbon dioxide is transferred from rocks directly into the atmosphere by volcanic eruptions. This process releases an average of roughly 0.1 GtC/yr. Although this flux is small compared with other fluxes, over millions of years it can lead to significant transfers of carbon into the atmosphere–land biosphere–ocean system.

Volcanic emissions of carbon dioxide are roughly balanced by a process known as *chemical weathering*, which removes about an equal amount of carbon from the atmosphere–land biosphere–ocean reservoir and transfers it back into rocks. Chemical weathering starts when carbon dioxide in the atmosphere dissolves into raindrops falling toward the surface through the process shown in Equation 5.3 (remember that carbon dioxide dissolves readily in water). Carbonic acid (H_2CO_3) is produced, which means that rain is slightly acidic (pH = 5.6).

When this acidic rain falls on rocks, both the physical impact of the rain drops and chemical reactions break the rock down. The chemical reaction is shown here:

$$CaSiO_3 + CO_2 \rightarrow CaCO_3 + SiO_2 \tag{5.5}$$

Note that this equation is a general description of the process of weathering, not the exact chemical reaction. Nonetheless, the essential message of Equation 5.5 is correct: The carbon dioxide molecule consumed in this reaction came from the atmosphere, via rainwater, and it is transferred into a molecule of calcium carbonate or $CaCO_3$, which is limestone and subsequently runs off with the rainwater and eventually reaches the ocean. The reaction also forms silicon dioxide (SiO_2), the primary component of sand, quartz, and glass.

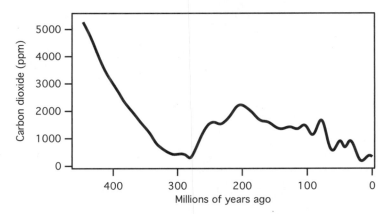

Fig. 5.4 Atmospheric carbon dioxide over the past half-billion years (based on Royer, 2006, Fig. 1).

Once in the ocean, the molecules of calcium carbonate are deposited through various mechanisms on the sea floor. Over millions of years, plate tectonics carries this calcium carbonate deep within the Earth, where high temperatures and pressures turn the rock into magma. Eventually, this magma is transferred back to the surface by volcanism, thereby releasing the carbon dioxide back to the atmosphere and completing the cycle.

Another pathway for carbon to move into the rock reservoir occurs when plants are rapidly buried in sediment before they can decay. This is, in fact, the same process that leads to a net production of oxygen (discussed earlier in the chapter). Once buried and subjected to the great heat and pressure found deep within the Earth, this dead plant material can be converted to fossil fuels, which humans extract and burn for energy – thereby returning the carbon to the atmosphere. We will explore fossil fuels in the next section.

A carbon atom will remain in the atmosphere–land biosphere–ocean system for approximately 40,000 GtC÷(0.1 GtC/yr) = 400,000 years before it is transferred into the rock reservoir. Given the large size of the rock reservoir and the relatively small rate of exchange between the rocks and the atmosphere, it takes many, many millions of years for a carbon atom to travel through the rock reservoir and reemerge into the atmosphere.

Figure 5.4 shows an estimate of atmospheric carbon dioxide over the past half-billion years. Four hundred million years ago, atmospheric carbon dioxide was more than 10 times higher than it is today, and since that time it has generally decreased, although there have been wide variations. Today's abundance, 390 ppm, is relatively low when compared to the geologic record. It is also worth pointing out that reconstructing ancient carbon dioxide abundances is difficult and none of the proxies are unambiguous. Thus, although Figure 5.4 shows our best guess for ancient carbon dioxide, there remains significant uncertainty concerning these estimates.

The variations in atmospheric carbon dioxide in Figure 5.4 are all due to variations in the rate of exchange of atmospheric carbon with the rock reservoir. This includes variations in the rate at which carbon dioxide is emitted from volcanoes; atmospheric carbon dioxide will increase, for example, during periods of extreme volcanism.

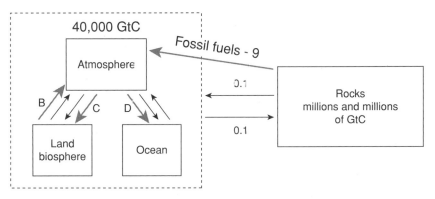

Fig. 5.5   Diagram of the carbon cycle as perturbed by humans. Gray arrows show net flows of carbon caused by human activities. Arrows B, C, and D represent deforestation, enhanced absorption of carbon by the land biosphere, and enhanced absorption of carbon by the ocean, respectively.

The movement of the continents is another factor. As continents move, patterns of rainfall can change, and new rock can be exposed to the atmosphere, both of which can change the rate of chemical weathering – and therefore the rate that carbon dioxide is removed from the atmosphere. For example, approximately 40 million years ago the Indian subcontinent collided with the Asian continent, forming the Himalayas and the adjacent Tibetan Plateau. Changing wind patterns brought heavy rainfall onto the expanse of the newly exposed rock, and the resultant chemical weathering drew down atmospheric carbon dioxide over a period of tens of millions of years.

5.4 How are humans perturbing the carbon cycle?

As Figure 5.4 shows, carbon dioxide can vary without any human activities. However, humans can also affect the carbon cycle. Figure 5.5 shows the perturbed carbon cycle, with the flows of carbon caused by human activities indicated as the gray lines. The main perturbation comes from the combustion of *fossil fuels* for energy. Fossil fuels were formed when plants that grew hundreds of millions of years ago were buried before the carbon in them could be released back into the atmosphere by respiration. Under high pressure and heat, applied over millions of years, the carbon in the plants was converted into the substances we know today as oil, coal, and natural gas. These fossil fuels are primarily carbon, with varying amounts of hydrogen and other trace species, such as sulfur.

When fossil fuels are burned, the net reaction is basically the same as the respiration reaction (Equation 5.2). The fossil fuel is combined with oxygen to produce energy and carbon dioxide. The energy is used to power our world, and the carbon dioxide is vented directly into the atmosphere.

Before humans discovered them, fossil fuels were safely sequestered in the rock reservoir. Under the processes of the natural carbon cycle, the carbon in these fossil fuels would be slowly released to the atmosphere through volcanic and other geological activity over many millions of years.

Humans, however, are extracting and combusting fossil fuels at a breathtaking pace – fast enough that most of the world's fossil fuels will be burned in just a few hundred years. The net result is the creation of an additional pathway for carbon from rocks to the atmosphere (the line marked "Fossil fuels" in Figure 5.5). And this pathway is large: In 2008, fossil fuel combustion released approximately 9 GtC to the atmosphere – roughly 90 times the natural flow rate of carbon from the rocks to the atmosphere.

Humans have also been chopping down large tracts of forest – a process known as *deforestation* – in order to use the land for other activities, such as agriculture or grazing livestock. Frequently, the forest is removed by burning it, which releases the carbon stored in trees and other plants to the atmosphere. Even just bulldozing the forest releases the carbon to the atmosphere, albeit more slowly. In addition, the soil often contains large amounts of carbon stored in the form of dead organic plant material. When the forest is removed, much of this carbon is subsequently released back into the atmosphere. Deforestation is an important source of carbon dioxide for the atmosphere, and estimates are that it contributed approximately 1.5 to 2 GtC to the atmosphere per year during the 2000s. This flux is shown in Figure 5.5 as Arrow B.

The impact of human activities on atmospheric carbon dioxide is shown in Figure 5.6. Figure 5.6a shows the abundance of carbon dioxide over the past 10,000 years. Over almost all of this time, the abundance remained in a narrow range, 260–280 ppm, but a sudden spike occurred in the past few hundred years. Figure 5.6b shows a close-up of the past 250 years. It shows that atmospheric carbon dioxide began rapidly increasing right at the beginning of the industrial revolution, when widespread economy-wide burning of fossil fuels began.

Figure 5.6b also shows that the rise in carbon dioxide is accelerating. Over the past 250 years, atmospheric carbon dioxide has increased by 100 ppm or so. The first 50 ppm of the increase took more than 200 years, until the 1970s, whereas the second 50 ppm took only 30 years. Figure 5.6c shows high-resolution measurements of the abundance of carbon dioxide in our atmosphere over the past 50 years. The yearly sawtooth pattern reflects the seasonal cycle in plant growth, which was discussed in Section 5.2.1. There is also a long-term increase in carbon dioxide, from 315 ppm in the late 1950s to 390 ppm in 2010. This increase is primarily due to fossil fuel combustion, with a smaller but important contribution from deforestation.

Figure 5.7 shows the year-to-year increase in atmospheric carbon dioxide. In the late 1950s, the increase in atmospheric carbon dioxide was less than 1 ppm/yr, while during the first decade of the 21st century, the increase was nearly 2 ppm/yr. This reflects the increasing rate of fossil fuel consumption over the past half-century.

We have good records of exactly how much fossil fuel is extracted and burned each year. In 2008, for example, the combustion of fossil fuels led to the emission of approximately 8.7 GtC to the atmosphere. Coal combustion contributed 40% of this, oil combustion contributed 36%, and most of the rest comes from natural gas (methane) combustion. From this, we can calculate how much atmospheric carbon dioxide should have gone up had all of it remained in the atmosphere. This is shown as the solid curve in Figure 5.7. Over the past 50 years, the increase in atmospheric carbon dioxide is approximately half of the total amount emitted to the atmosphere.

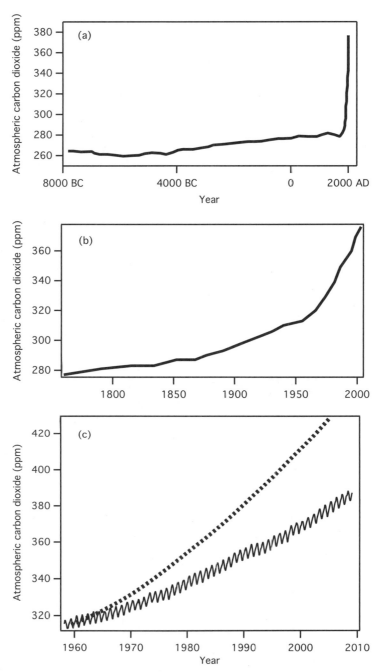

Fig. 5.6   Abundance of carbon dioxide in our atmosphere over the past (a) 10,000 years, (b) 250 years, and (c) 50 years
(sawtooth line). Panel c also shows (dotted curve) the annual average carbon dioxide abundance if all of the
carbon dioxide emitted by human activities since 1959 had remained in the atmosphere. (Panels a and b are
adapted from Figure SPM.1 of IPCC, 2007a; measurements in panel c are obtained from the NOAA Earth System
Research Laboratory/Global Monitoring Division, at http://www.esrl.noaa.gov/gmd/ccgg/trends/;
the 100%-retained line is calculated from Denman et al., 2007, Fig. 7.4.)

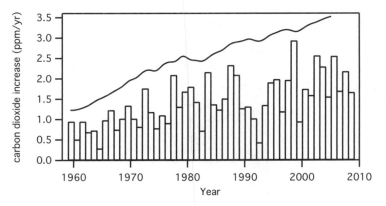

Fig. 5.7 Bars show the observed year-to-year increase in atmospheric carbon dioxide. The solid line shows what the annual increase would have been had 100% of the carbon dioxide emissions remained in the atmosphere (bars are provided by the NOAA Earth System Research Laboratory/Global Monitoring Division, at http://www.esrl.noaa.gov/gmd/ccgg/trends/; the solid line is from Denman et al., 2007, Fig. 7.4).

Thus, in 2008, humans released 10 GtC into the atmosphere, and the amount of carbon in the atmosphere increased by 5 GtC. Of the 5 GtC that were removed, roughly half, 2.5 GtC, went into the ocean. This is indicated as Arrow D in Figure 5.5. The other half is believed to have gone into the land biosphere, although there is considerable scientific debate about exactly what part of the land biosphere is absorbing the carbon. This enhanced land sink is represented by Arrow C in Figure 5.5.

There is relatively large year-to-year variability in the fraction of carbon dioxide that is retained in the atmosphere. For example, atmospheric carbon dioxide increased by nearly 3 ppm in 1998, but less than 1 ppm the very next year – even though emissions were roughly the same. This is mainly due to variations in the climate due to El Niño events. These events can affect the carbon cycle several ways, such as by changing the absorption of carbon dioxide by the oceans, or by modulating the occurrence of forest fires. It turns out that the occurrence of El Niños and La Niñas explains much of the year-to-year variations in the fraction of emitted carbon dioxide that remains in the atmosphere.

If the land biosphere and ocean were not taking up some of the carbon we were emitting, then atmospheric carbon dioxide would be much higher today than it actually is. Figure 5.6c shows what the long-term time series of atmospheric carbon dioxide would be if it all emissions from human activities since 1959 had remained in the atmosphere – it shows that atmospheric carbon dioxide would be near 430 ppm in 2005, approximately 50 ppm higher than the actual abundance. The climate would consequently be warmer and changing even more rapidly. Thus, the land biosphere and ocean are doing us a huge favor by absorbing significant amounts of carbon emitted by humans.

An emerging concern of some scientists is whether the oceans and land biosphere will take up as much carbon in the future as they presently are. It is unknown when or if we will reach a saturation point at which the reservoirs slow down or even cease their uptake. If that happens, then a higher fraction of subsequent emissions will remain in the atmosphere, and the abundance of carbon dioxide in our atmosphere will grow more rapidly. This leads to yet another worry by climate scientists: that climate change itself may alter the carbon cycle. In the next chapter, we will explore how such carbon cycle feedbacks may amplify climate change.

5.5 Some commonly asked questions about the carbon cycle

Because of its central role in the climate change problem, climate skeptics occasionally challenge the claim that the recent increase in carbon dioxide is due to human activities. In this section I address this argument.

How do we know that combustion of fossil fuels is responsible for the increase in carbon dioxide, instead of nonhuman sources such as volcanoes or plants?

This is a reasonable question. After all, the amount of carbon dioxide absorbed by plants during the year and balanced by plant decay (approximately 100 GtC/yr) is much larger than human emissions (currently at 10 GtC/yr or so) – ditto for the ocean fluxes. So it may seem reasonable that the increase in atmospheric carbon dioxide in the atmosphere might be driven by a slight excess of plant respiration over photosynthesis, or a slight excess flux of carbon dioxide out of the ocean. Similarly, we know that volcanoes emit carbon dioxide, and that over millions of years, volcanoes are a primary source of carbon dioxide in the atmosphere. So maybe the increase in atmospheric carbon dioxide is due to enhanced volcanic activity.

There are, however, several independent lines of evidence that unanimously agree that fossil fuel combustion is the dominant reason for the increase in atmospheric carbon dioxide over the past few centuries. First, Figure 5.7 shows that, for the past half-century, each year's increase in carbon dioxide in the atmosphere has been roughly half of what humans released into the atmosphere in that same year. Thus, when humans were emitting smaller amounts of carbon dioxide in the 1960s, atmospheric carbon dioxide was increasing at a slower rate than when humans were dumping large amounts of carbon dioxide in the atmosphere, as we are today. Moreover, the increase in the atmospheric carbon dioxide in a year never exceeds what humans emitted during that year. If the source of carbon dioxide emissions were nonhuman, it seems unlikely that it would track human emissions of carbon dioxide so closely.

Second, the carbon dioxide can be chemically "fingerprinted" to show that it comes from fossil fuels. The method is based on *isotopes* of carbon. All carbon atoms have six protons, but carbon's isotopes have different numbers of neutrons. The most abundant isotope is carbon-12, containing six neutrons to go with the six protons, and which makes up roughly 99% of the carbon on Earth. Carbon-13, with seven neutrons, makes up 1% of the carbon or so, and approximately one carbon atom out of a trillion is carbon-14, which has eight neutrons.

The chemical properties of an atom are for the most part set by the number of protons, so isotopes tend to have very similar chemical properties. The chemistry, though, is not identical. Plants, for example, preferentially absorb carbon-12 when growing. Because fossil fuels are derived from plants, they should reflect this pref- erence for carbon-12 over carbon-13. And the carbon dioxide produced from fossil fuel combustion will also reflect the preference for carbon-12 over carbon-13.

Scientists can measure the amount of carbon-12 and carbon-13 in atmospheric carbon dioxide, and those measurements show that the increase in atmospheric carbon dioxide in Figure 5.6b is caused by carbon that is depleted in carbon-13 – such as that which comes from plants. This allows us to rule out sources such as volcanoes or the ocean.

So we know that the increase in atmospheric carbon dioxide is coming from plants, but is it coming from plants that died hundreds of millions of years ago (i.e., fossil fuels) or plants of today? In order to make that determination, we turn to carbon-14. Carbon-14 is produced in the atmosphere when a neutron created by a cosmic ray hits the nucleus of an atom of nitrogen-14. The nucleus absorbs the neutron and ejects a proton, thereby transforming itself into carbon-14. Carbon-14 atoms are incorporated into molecules of carbon dioxide and are then absorbed by plants and incorporated into plant material. If you walk outside and pull a leaf off a tree, approximately one out of every trillion carbon atoms in that leaf would be carbon-14.

Carbon-14 is known as radiocarbon because it is radioactive. That means its nucleus is unstable and converts back to nitrogen-14 with a half-life of approximately 6,000 years (this means that, after 6,000 years, half of the carbon-14 has converted back to nitrogen-14). So imagine a cotton plant that grew 6,000 years ago. As it grew, the plant absorbed carbon dioxide containing carbon-14 from the atmosphere, and the carbon-14 was incorporated into the plant. Immediately after the plant was picked, it would have the same proportion of carbon-14 as any plant, and so would the cotton produced from it. Because it was no longer alive, however, it stopped absorbing carbon dioxide from the atmosphere. Over time, the amount of carbon-14 in the cotton slowly decreased as it was converted back to nitrogen-14.

Now imagine that modern-day archaeologists find a blanket made of this cotton and want to know how old it is. To do this, they measure the proportion of carbon-14 in the blanket and find that it has half the carbon-14 of a living plant. With a half-life of 6,000 years, the archaeologists conclude that the blanket is 6,000 years old. If they found that it had one fourth of the carbon-14 of a living plant, then it would be 12,000 years old. This process is known as *radiocarbon dating*.

Now let's turn our attention to fossil fuels. As we learned earlier, fossil fuels are produced when plant matter is buried for millions of years. After millions of years of being underground, all of the carbon-14 has converted back to nitrogen-14. Thus, fossil fuels contain essentially no carbon-14; this condition is known as *radiocarbon dead*. When the fossil fuels are burned, the carbon dioxide produced also has no carbon-14 in it.

Thus, carbon dioxide from fossil fuels has no carbon-14 whereas carbon dioxide from the land biosphere is relatively rich in carbon-14. Measurements indicate that the carbon dioxide being added to the atmosphere is indeed radiocarbon dead, showing that it is coming from long-dead plants – fossil fuels – and not modern plants.

Putting all of the evidence together, along with an absence of any counterevidence, we see that there is no question that human activities are increasing the amount of carbon dioxide in the atmosphere. As my colleague John Nielsen-Gammon puts it, not only can we see the smoking gun, but the smoke is a chemical match to the gunpowder.

Why focus on carbon dioxide emitted during the combustion of fossil fuels when plants and animals emit far more carbon dioxide to the atmosphere?

Humans, animals, bacteria, and plants do indeed emit enormous amounts of carbon dioxide to the atmosphere – the land biosphere emits 100 GtC/yr, compared with emissions from fossil fuels of less than 10 GtC/yr. So why should we care about carbon dioxide from fossil fuels? To understand the answer, you need to understand the difference between carbon dioxide coming from fossil fuel combustion and from respiration by living organisms.

Let's begin by imagining that you plant a carrot seed, and over the next few months this seed grows into a carrot. As described by Equation 5.1, the plant grows by absorbing carbon dioxide directly from the atmosphere. So the carrot contains carbon that, just a few months ago, was floating in the atmosphere as carbon dioxide.

Now let's assume that the carrot is eaten by a goat. The goat metabolizes the carrot (in a manner approximately following Equation 5.2), which produces energy that is used to power the goat's vital functions. The carbon dioxide is also produced, and it is exhaled back into the atmosphere.

Thus, when an animal exhales carbon dioxide, it is releasing back into the atmosphere the carbon dioxide that was in the atmosphere just a few months before. Although this can lead to seasonal variations in carbon dioxide, as shown in Figure 5.1, it does not cause long-term increases in carbon dioxide. Figure 5.6 confirms this by showing basically no change in carbon dioxide over the past 10,000 years, before about 1750. Over this time we know that humans, plants, and animals were certainly releasing carbon dioxide to the atmosphere, yet atmospheric carbon dioxide remained approximately constant.

In contrast, when you burn fossil fuels, you are releasing to the atmosphere the carbon dioxide that had been safely sequestered in rocks (e.g., Figure 5.5) for hundreds of millions of years. This is a net addition to the atmosphere, so it does cause a long-term increase in carbon dioxide. Figure 5.6 confirms this: The increase in atmospheric carbon dioxide started with the industrial revolution, when society-wide burning of fossil fuels began.

5.6 Methane

Most discussions of the carbon cycle focus on the cycling of atmospheric carbon dioxide. However, methane is another crucial carbon-containing gas. Although the atmospheric abundance of methane is much smaller than that of carbon dioxide (1.8 ppm or so in 2010), on a per molecule basis methane is roughly 20 times more powerful of a greenhouse gas than carbon dioxide.

Atmospheric methane has both human and natural sources. The most important human source is the raising of livestock. Cattle, as well as goats and sheep, are ruminants, and these animals produce methane in their guts during the digestion of

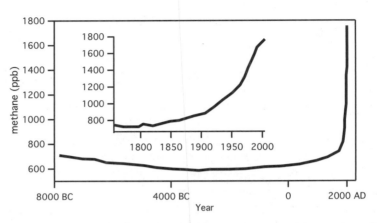

Fig. 5.8 Methane abundance over the past 10,000 years. The inset plot shows a close-up of the past 250 years (adapted from the IPCC Fourth Assessment Report Summary for Policymakers, 2007, Fig. SPM.1).

food. This methane is eventually released to the atmosphere (out of both ends of the animals), and it totals about 80 million tons per year. The next biggest source is emissions from rice paddies. In the warm and wet environment of a flooded rice field, bacteria in the soil efficiently produce methane, the vast majority of which is then released to the atmosphere. Such emissions total about 60 million tons per year. Bacterial processes in landfills and other waste repositories are today contributing about 60 million tons per year, and the release of methane from the petrochemical industry adds another 60 million tons per year. Release of geologic methane from coal mines adds 50 million tons per year to the atmosphere. Finally, burning of forest and other biomass primarily produces carbon dioxide, but it also produces methane if the combustion temperature is sufficiently low (e.g., a smoldering fire). Such biomass burning adds 50 million tons to the atmosphere each year. In total, annual emissions of methane associated with human activities in the 2000s were approximately 350 million tons.

The annual total emission of methane from natural sources is roughly 200 million tons per year. Approximately three fourths of these natural emissions are from natural wetlands, which produce methane the same way that flooded rice paddies do. Minor contributions come from termites (20 million tons per year), emissions from the ocean (10 million tons per year), natural geological sources such as seepage in sedimentary basins and geothermal–volcanic emissions (10 million tons per year), and wild animals (a few million tons per year).

Thus, 35% of the methane emissions are natural, and 65% are due to human activities. Looking at it another way, we see that 70% of the global total methane emissions come from bacterial fermentation of organic material. These sources include wetlands, rice agriculture, livestock, landfills, forests, oceans, and termites. The other 30% of the emissions are nonbiogenic, and they include emissions from fossil fuel production, biomass burning, and geological sources.

Figure 5.8 shows the atmospheric abundances of methane began rising 200 years ago, which scientists have attributed to human activities. As we will see in Chapter 6, these emissions are enough to make methane a significant contributor to the warming we are experiencing – methane's contribution to global warming is approximately one fourth of the contribution of carbon dioxide. As a result, reductions of methane emissions are frequently included in plans to address climate change.

Methane is removed from the atmosphere by oxidation, which follows the following schematic reaction:

$$CH_4 + 2O_2 \rightarrow CO_2 + 2H_2O$$

On average, a molecule of methane is destroyed by this reaction 10 years after it was emitted. If we stopped emitting methane today, within a few decades all of the human-emitted methane would be gone, and the atmospheric abundance would be back down to pre-industrial amounts. This is quite different from carbon dioxide, which can stay in the atmosphere for centuries or millennia. We will return to this topic in Chapter 8.

5.7 Chapter summary

- Only a few components of our atmosphere are greenhouse gases, which absorb infrared photons. The three most important are (in order) water vapor, carbon dioxide, and methane. Nitrogen, oxygen, and argon, which make up approximately 99.9% of the dry atmosphere, are not greenhouse gases.
- The carbon cycle describes how carbon cycles through its primary reservoirs: the atmosphere (containing 100 GtC), land biosphere (2,000 GtC), ocean (1,000 GtC in the mixed layer and 38,000 GtC in the deep ocean), and rocks (millions and millions of gigatonnes of carbon).
- The atmosphere exchanges carbon with the land biosphere through photosynthesis and respiration. The atmosphere exchanges carbon with the ocean when carbon dioxide dissolves into the ocean. Once in the ocean, the carbon dioxide is converted to carbonic acid and other chemicals. Over the course of several centuries, a carbon atom added to the atmosphere will cycle through all of the other reservoirs and return to the atmosphere.
- The atmosphere–land biosphere–ocean system also exchanges carbon with the rocks containing carbon through volcanism and chemical weathering. This exchange is normally extremely slow.
- Humans are perturbing the carbon cycle by extracting and burning fossil fuels. The result is the creation of a new pathway for carbon from rocks to the atmosphere. In 2008, fossil fuel combustion released 9 GtC to the atmosphere from the rock reservoir, which is approximately 90 times the amount released from the rocks naturally. Deforestation is another important human source, releasing 1.5–2 GtC/yr from the land-biosphere and into the atmosphere during the 2000s.
- Roughly half of the carbon dioxide emitted to the atmosphere is removed within 1 year or so by transport into the land biosphere and ocean. The other half remains

in the atmosphere for much longer and has caused an increase in the atmospheric abundance from approximately 280 ppm in 1750, before the industrial revolution, to 390 ppm in 2010.

- Methane is another important greenhouse gas – each molecule of methane has the warming power of approximately 20 carbon dioxide molecules. Two thirds or so of the methane emissions are due to human activities, whereas one third is from natural sources. Human emissions have caused an increase in the atmospheric abundance from approximately 0.8 ppm in 1750, before the industrial revolution, to 1.8 ppm in 2010.

Additional reading

D. Archer, *The Global Carbon Cycle* (Princeton, NJ: Princeton University Press, 2010). This is a short and focused textbook on carbon cycle science. If you read it, you will know more than just about anyone else about how carbon moves around the planet.

E. Roston, *The Carbon Age: How Life's Core Element Has Become Civilization's Greatest Threat* (New York: Walker, 2009). This is a fun and easy-to-read book about carbon and the immense role it plays in our lives.

Terms

Carbon cycle
Carbonic acid
Chemical weathering
Deep ocean
Deforestation
Fossil fuels
Greenhouse gas
Halocarbons
Isotopes
Mixed layer
Parts per million
Photosynthesis
Radiocarbon dating
Radiocarbon dead
Respiration
Time scale
Turnover time

Problems

1. a) Explain the processes that transfer carbon from the atmosphere to the land, and from the land to the atmosphere. What are the chemical reactions that describe these processes?

   b) How do these processes interact to produce the "sawtooth" annual cycle in the atmospheric abundance of CO_2 shown in Figures 5.1 and 5.6?

2. A letter to the editor of the *Austin American-Statesman*, published on December 23, 2009, asks this question: "The trillion-dollar question that Copenhagen has not answered [is this]: Because carbon dioxide molecules are all identical, why is it that carbon dioxide from carbonated beverages, pets, cattle, farm animals, and humans, east, dry ice, fireplaces, charcoal grills, campfires, wildfires, alcohol and ethanol is good, and carbon dioxide from fossil fuel is bad? Can anyone in the United States answer this question?" What's your answer?

3. Your aunt asks you how we know that humans are responsible for the increase in atmospheric CO_2. Couldn't it be due to volcanoes? Or could it be coming from plants? What do you tell her?

4. Explain how isotopes help us identify human activities as the reason atmospheric carbon dioxide is increasing.

5. Your grandmother asks you to explain how humans are modifying the carbon cycle. What do you tell her?

6. Explain how "chemical weathering" removes CO_2 from the atmosphere. What is the weathering chemical reaction? Can this process play an important role in counteracting the increase in atmospheric carbon dioxide caused by humans?

7. Of the carbon dioxide humans add to the climate, approximately half is removed within a year. Where does it go? How would it affect the climate if, all of the sudden, all of the carbon dioxide we emit stayed in the atmosphere?

8. Why is rain naturally acidic? What then, does the term *acid rain* refer to? (Acid rain is not covered in the chapter, so you'll have to do some outside research on it.)

6

Forcing, feedbacks, and climate sensitivity

In Chapter 4, we showed that the temperature of a planet is a function of the solar constant, the albedo of the planet, and the composition of the atmosphere (Equation 4.5). If that were all there was to climate change, this would be a pretty short book. Unfortunately (or fortunately, depending on how you look at it), there is a lot more interesting physics that we have to consider to fully understand the problem of modern climate change.

6.1 Time lags in the climate system

In Chapter 3, we talked about how an oven cooks food, and how it takes several hours for a turkey put into an oven to heat up. One can think of the time lag between putting the turkey in the oven and its being cooked as a manifestation of the turkey's *thermal inertia* – just like the inertia that keeps your car from instantaneously stopping when you hit the brakes.

In an analogous way, the temperature of the Earth does not instantaneously respond to additions of greenhouse gas to the atmosphere. To understand this better, let's consider a planet with no atmosphere that is in equilibrium (energy in equals energy out, or $E_{in} = E_{out}$, for the entire planet). Given $E_{in} = 238$ W/m^2 (the value for the Earth), we calculated in Chapter 4 that the planet's surface temperature would be 255 K (Figure 6.1a).

Now let's imagine that a one-layer atmosphere is instantly added to the planet. What are the fluxes the instant after the layer is added? The temperature of the surface has not changed (the temperature of the surface cannot change instantly any more than a turkey can cook instantly or a car can stop instantly), so it remains 255 K. This in turn means it is emitting exactly the same as it was before the atmosphere was added, 238 W/m^2. The atmosphere, however, is now absorbing all of the photons coming from the surface. Half of the absorbed energy is reemitted upward to space, and half is reemitted downward back to surface.[1] This is shown in Figure 6.1b.

This means that energy out for the planet has been cut by half, to 119 W/m^2. Given that energy in remains the same, 238 W/m^2, this tells us that energy in exceeds energy out for the planet, so the planet must be warming. Moreover, the energy that is no longer escaping to space, also 119 W/m^2, has been redirected by the atmosphere

---

[1] I am implicitly assuming here that energy in always equals energy out for the atmosphere. In more technical terms, I am assuming the heat capacity of the atmosphere is zero. This is not a bad assumption because, while not zero, the heat capacity of the atmosphere is indeed much smaller than the heat capacity of the rest of the planet.

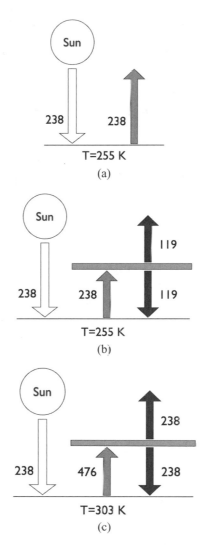

Fig. 6.1 A schematic of energy fluxes on a planet (a) with no atmosphere, (b) the instant after a one-layer atmosphere is added to the planet, and (c) after the climate reaches its new equilibrium.

back toward the surface. As a result, energy in for the surface is now $E_{in} = 238 + 119 = 357$ W/m^2, which exceeds energy out of $E_{out} = 238$ W/m^2, so the surface is also warming. As the surface and rest of the planet warms, energy out increases. Eventually, the surface and atmosphere warm enough that $E_{in} = E_{out}$ and the planet is again in energy balance, as shown in Figure 6.1c.

As we discussed previously, adding greenhouse gases to the atmosphere is the same as adding layers. Thus, what we have really worked out here is what happens when the atmospheric abundance of greenhouse gases increases. The greenhouse gases intercept some of the energy escaping to space and redirect it back toward the surface. In this way, greenhouse gases both reduce energy out for the planet and increase energy in for the surface, thereby warming the planet. The planet warms until energy out again balances with energy in for both the surface and the planet as a whole.

This warming does not happen instantly. For the Earth, it takes a few decades to warm the climate in response to an increase in greenhouse gases because 70% of the Earth is covered by ocean, and water can hold a huge amount of heat (something you know if you've ever waited for a hot tub to heat up or tried to boil a big pot of water). Thus, the planetary energy imbalance, energy in minus energy out, or $E_{in} - E_{out}$, must persist for decades in order to accumulate enough energy in the ocean to significantly warm it up.

At present, energy in for the Earth exceeds energy out by approximately 0.9 W/m^2, primarily as a result of the addition of greenhouse gases to our atmosphere by human activities. With this imbalance, Earth needs to warm up by approximately half a degree in order for energy out to again equal energy in. Thus, if we stopped emitting greenhouse gases today, *we would still experience approximately a half a degree warming over the next few decades.*[2] This is often referred to as *committed warming* or described more informally as warming "in the pipeline." We've already paid for this warming with emissions over the past few decades, but the warming has not yet arrived. Nevertheless, it is coming – and there's basically nothing we can do to avoid it.

The lag between the emission of carbon dioxide and the associated warming has important implications for the policy debate over climate change. Because warming from today's emissions will not be fully realized until the middle of the 21st century, the benefits from reductions in emissions today, which may be costly, will also not be fully realized for several decades. This means that reducing emissions requires today's population to pay costs that may primarily benefit future generations. When we explore policy options later in the book, we will return to this problem.

6.2 Radiative forcing

Figure 6.2 plots energy out to space for the planet shown in Figure 6.1. The period before Year 0 corresponds to Figure 6.1a, when the planet had no atmosphere and it was in energy balance. At Year 0, the atmosphere is instantaneously added, and energy out drops immediately to 119 W/m^2. This corresponds to Figure 6.1b. As the surface heats up in response to the warming from the atmosphere, energy out increases, eventually reaching 238 W/m^2 – and the planet is once again in energy balance, corresponding to Figure 6.1c.

This leads us to one of the most important concepts in climate science: *radiative forcing*. Radiative forcing (often abbreviated RF) is the change in $E_{in} - E_{out}$ for the planet as a result of some change imposed on the planet *before the temperature of the planet has adjusted in response*[3]:

$$RF = \Delta(E_{in} - E_{out}) = \Delta E_{in} - \Delta E_{out} \qquad (6.1)$$

[2] The exact amount of warming depends on what we assume for emissions of non-greenhouse-gas forcers, such as that from aerosols.

[3] For somewhat technical reasons, radiative forcing is actually defined as the change in energy balance at the tropopause, but without allowing the tropospheric and surface temperatures to adjust. We ignore this technical point here.

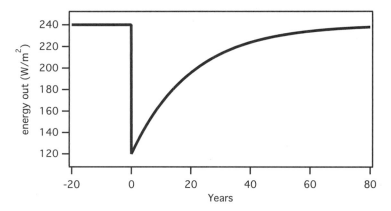

Fig. 6.2 Plot of energy out for the planet shown in Figure 6.1. The atmosphere is added instantaneously in Year 0. This planet has a heat capacity similar to the Earth's.

In the example just given, ΔE_{out} is -120 W/m^2; that's the drop in E_{out} the instant after the atmosphere is added, and before the warming temperature has caused E_{out} to increase. Note that ΔE_{in} is zero because energy in does not change when an atmospheric layer is added. Thus, the radiative forcing of adding a one-layer atmosphere is $0 - (-120) = +120$ W/m^2. Positive radiative forcings correspond to changes that warm the climate, whereas negative ones correspond to changes that cool the climate.

An example: What's the radiative forcing of a 5% increase in solar constant for the Earth that occurs over 100 years?

Let's begin by calculating ΔE_{in}, the change in energy in. From Chapter 4, we know that $E_{in} = S(1 - \alpha)/4$, which for the Earth is 238 W/m^2. If the solar constant S increased by 5%, then S would increase to $1{,}360(1.05) = 1428$ W/m^2. For this new value of the solar constant, $E_{in} = 250$ W/m^2. Thus, E_{in} has increased from 238 W/m^2 to 250 W/m^2, so $\Delta E_{in} = +12$ W/m^2.

How does ΔE_{out} change when the solar constant changes? E_{out} is determined entirely by atmospheric composition (i.e., number of layers) and temperature. Atmospheric composition is not changing in this example, and radiative forcing is defined to be the response to an instantaneous change, *before the temperature of the planet has adjusted to the change*. Thus, E_{out} does not change and therefore $\Delta E_{out} = 0$. Putting it together (Equation 6.1), we see that the radiative forcing for this change in the solar constant is $+12$ W/m^2.

Note that the fact that the change occurred over the course of 100 years does not enter into the calculation. All radiative forcing calculations are done under the assumption that the change occurs instantly and the climate is not allowed to respond to the change.

The imposition of a radiative forcing on a planet, such as a change in solar constant, will take the planet out of energy balance – so that E_{in} and E_{out} are no longer equal to each other. In response, the temperature of the planet will adjust so that E_{in} once

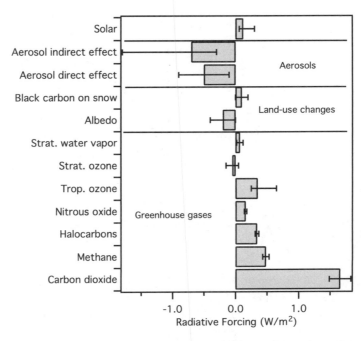

Fig. 6.3 Radiative forcing caused by changes in the climate between 1750 and 2005. The error bars indicate the uncertainty of the estimate (based on IPCC, 2007a, Figure SPM.2).

again equals E_{out}. In the case of an increasing solar constant, the planet will warm, increasing E_{out}, until E_{in} and E_{out} are once again in balance.

Thus, radiative forcing is a quantitative measure of how much some climate perturbation (e.g., an increase in solar constant, or increase in greenhouse gases) will change the climate. The advantage of using radiative forcing is that it allows us to express diverse changes to the climate system by using a common metric. For example, it allows us to compare the climate-changing effect of a 100-ppm increase in carbon dioxide to a 1% increase in the solar constant. By comparing the radiative forcing of these two changes, we could determine which one would warm the planet more. Similarly, radiative forcing of $+1$ W/m^2 will produce similar warming of the climate, regardless of whether that change was caused by a brightening of the Sun, an increase in carbon dioxide, an increase in methane, or some other change.

Figure 6.3 shows the radiative forcing of the various factors that have influenced our climate over the past few centuries. In much the same way that temperature anomalies are the change in temperature from a reference period (see Chapter 2), radiative forcings are generally calculated as a change from a reference climate. In Figure 6.3, the values plotted are radiative forcing caused by changes since 1750, which is considered the pre-industrial value. In the rest of this section, I describe each one of these factors.

6.2.1 Trace gases

The atmospheric abundance of carbon dioxide increased by 100 ppm between 1750 and 2005. If we instantaneously increase carbon dioxide from 280 to 380 ppm, energy out will instantaneously decrease by 1.66 W/m^2 (energy in does not significantly change). Thus, the radiative forcing of carbon dioxide is $+1.66$ W/m^2. Increases in methane, nitrous oxide, and the halocarbons between 1750 and 2005 produce radiative forcings of $+0.48$, $+0.16$, and $+0.34$ W/m^2, respectively, for a total of $+0.98$ W/m^2.

Ozone in the lower atmosphere is both a greenhouse gas and one of the primary components of photochemical smog. As the world has become more industrialized, lower atmospheric ozone has increased along with the other components of air pollution. This increase contributes a positive radiative forcing of $+0.35$ W/m^2. Ozone in the stratosphere, in contrast, has been declining as a result of ozone depletion. This contributes a negative radiative forcing of -0.05 W/m^2.

Next is stratospheric water vapor. An important source of stratospheric water vapor is the transport of methane into the stratosphere followed by oxidation, which has this net reaction:

$$CH_4 + 2\,O_2 \rightarrow 2\,H_2O + CO_2 \tag{6.2}$$

The increase in methane over the past two centuries has therefore increased stratospheric water, which has led to a positive radiative forcing of $+0.07$ W/m^2. We'll consider lower-atmospheric water vapor in the section on feedbacks.

Thus, although carbon dioxide is the single most important greenhouse gas emitted by human activities, it is not the only important one. In fact, the combined radiative forcing from the other greenhouse gases ($+1.35$ W/m^2) is comparable to the radiative forcing from carbon dioxide alone ($+1.66$ W/m^2). As we will see in the last chapter of this book, this has important implications for policies to address climate change.

6.2.2 Aerosols

Aerosols are particles so small that they do not fall under the force of gravity, but remain suspended in the atmosphere for days or weeks. Like clouds, aerosols can interact both with sunlight that is falling on the planet and on infrared radiation that is being emitted by the surface and atmosphere – and thereby alter the climate. There are several types of aerosols, and their composition determines how they interact with sunlight and infrared radiation.

When fossil fuels containing sulfur impurities are burned, the sulfur is released to the atmosphere with the other products of combustion. Once in the atmosphere, the sulfur gases react with other atmospheric constituents to form small liquid droplets, known as sulfate aerosols. Sulfur is also released into the atmosphere during biomass burning and from natural processes in the ocean.

Sulfate aerosols are highly reflective and reflect incoming solar radiation back to space, so their net effect is to cool the climate. As a result of increases in fossil fuel use over the past two centuries, the abundance of sulfate aerosols has steadily increased with time, providing a negative radiative forcing of -0.4 W/m^2.

Such sulfate aerosols all occur in the lower atmosphere, so their lifetime is short – it takes just a few weeks before the aerosols are either washed out of the atmosphere by rain or fall to the ground. This means that the radiative forcing the Earth is experiencing at any given time from these aerosols is due entirely to emissions of sulfur that occurred in the past month or two.

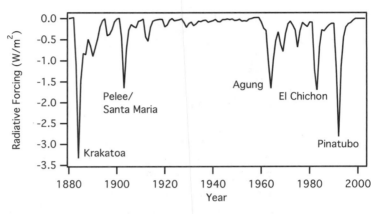

Fig. 6.4   Radiative forcing from volcanoes (data obtained from NASA GISS; see http://data.giss.nasa.gov/modelforce/RadF.txt).

Another important – but episodic – source of sulfur gases for the atmosphere is volcanic eruptions. Volcanoes emit enormous amounts of sulfur gas, and energetic eruptions can inject it directly into the stratosphere. Aerosols in the stratosphere can remain there for several years – much longer than an aerosol resides in the lower atmosphere. This long life-time, combined with the massive amounts of sulfur released, means that a single volcano can produce a negative radiative forcing of several watts per square meter that lasts for several years after the eruption (Figure 6.4).

This negative radiative forcing can lead to a significant cooling of the climate following an eruption. In 1816, for example, after three major eruptions in three years, the United States and Europe experienced the "year without a summer," in which snow fell in Vermont in June and heavy summer frosts caused crop failures and widespread food shortages. When that summer was followed by a winter so cold that the mercury in thermometers froze (this happens at $-40\,°C$), many residents fled the Northeast United States and moved south.

After a few years, stratospheric aerosols fall out of the stratosphere and the climate warms back up. Combined with the fact that such massive volcanic eruptions occur infrequently (as we can see in Figure 6.4), the long-term impact of volcanoes on the climate has been relatively small over the past few centuries.

Black carbon aerosols, such as soot, are another important aerosol type. This type of aerosol is produced by incomplete combustion, such as a smoldering fire or by two-stroke gasoline engines, so they are generally of human origin. Because they are dark, they absorb solar radiation and decrease the planet's albedo, tending to heat the planet. Over the past few centuries, black carbon aerosol abundance has increased as more people burn more stuff, leading to a positive radiative forcing of $+0.2\ W/m^2$. Much like sulfate aerosols, these black carbon aerosols have atmospheric lifetimes of a few weeks.

Another type of aerosol is mineral dust. Most of this dust comes from natural processes, such as dust that is picked up off the world's deserts by strong winds (Figure 6.5). Approximately 20% of mineral dust comes from anthropogenic sources – mainly agricultural practices (e.g., harvesting, plowing, overgrazing), changes in surface water features (e.g., drying out of lakes such as the Aral Sea and Lake Owens)

Fig. 6.5   Image of a strong temperate cyclone over China, pushing a wall of dust as it moved. The image was captured in early April of 2001 by the Moderate Resolution Imaging Spectroradiometer on NASA's Terra satellite (image obtained from the Earth Observatory; see http://earthobservatory.nasa.gov/IOTD/view.php?id=8341). (See Color Plate 6.5.)

and industrial practices (e.g., cement production, transport). The net effect of dust is to cool the planet, and human activities have contributed a negative radiative forcing of -0.1 W/m^2. Like other types of aerosols, these dust aerosols have atmospheric lifetimes of a few weeks.

Combining all types of aerosols (those discussed above and several not discussed), the *direct radiative effect* of aerosols is to cool the climate, with an estimated negative radiative forcing of -0.5 W/m^2. However, aerosols also have an *indirect effect* on the climate, whereby aerosols influence the climate by altering clouds. There are several ways that this can occur. One of the best-understood mechanisms is by altering the number of particles in a cloud. Cloud particles generally form when water condenses onto a *cloud-condensation nuclei* or CCN, which are small solid or liquid aerosols that are hydrophilic, meaning that they attract water. Thus, the number of CCN in an air mass influences how many cloud particles are found in a cloud.

If you add aerosols to a cloud, then you will increase the number of CCN – and therefore the number of cloud droplets making up the cloud. At the same time, the total liquid water contained in the cloud is fixed. Thus, the increase in the number of droplets means that each droplet has less water and is therefore smaller. This is akin to cutting a pie into more slices: The total amount of pie is fixed, so more slices means that each slice must be smaller.

It turns out that a cloud containing a larger number of smaller droplets is more reflective. A familiar example of this can be seen in your kitchen in the difference between regular table sugar and powdered sugar. Chemically, the two substances are identical, but normal table sugar is made up of comparatively big particles, whereas

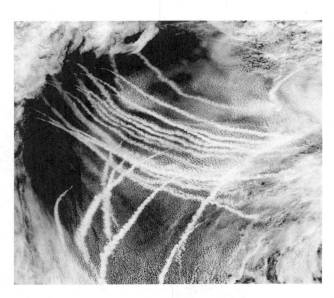

Fig. 6.6 Ship tracks in clouds off of the West Coast of the United States (image obtained from the Earth Observatory; see http://earthobservatory.nasa.gov/IOTD/view.php?id=37455). (See Color Plate 6.6.)

powdered sugar is made up of extremely small particles. Because smaller particles tend to be more reflective, the pile of powdered sugar is a brighter white than a pile of table sugar.

Thus, if one adds aerosols to a cloud, the cloud gets brighter and more reflective. This can be seen in what are called *ship tracks*. The exhaust from diesel engines contains fine particulates that can serve as CCN. As ships steam across the ocean, these fine aerosol particles from their engines are transported by the winds into low-level clouds, leading to increases in numbers of droplets and brighter clouds. From a satellite (Figure 6.6), lines of bright clouds trace out the paths of these ships.

This effect on cloud particle size is just one effect aerosols have on clouds. By making the cloud particles smaller, aerosols slow down the coagulation process whereby cloud droplets combine to form raindrops. This in turn means that the clouds don't rain, so they last longer. Aerosols can also change the height of the cloud, as well as the phase (ice vs. liquid). And addition of black carbon to clouds can lead to local warming that can cause clouds to evaporate.

Considering all the effects of aerosols on clouds, scientists estimate that the indirect aerosol effect produces a negative radiative forcing of -0.7 W/m^2. Together with the direct effect, aerosols therefore produce a negative radiative forcing of roughly -1.2 W/m^2. However, as the error bars in Figure 6.3 show, the indirect effect of aerosols is a highly uncertain number.

Because aerosols last only a few weeks in the atmosphere before they are removed, aerosols do not have time to become well mixed throughout the atmosphere (which takes a year or so). As a result, the distribution of aerosols in our atmosphere is highly variable, with most aerosols found near their sources. Figure 6.7 shows their distribution, and from this we can infer the major sources of aerosols across the globe: Saharan dust that is blown westward from North Africa into the Atlantic, aerosol from

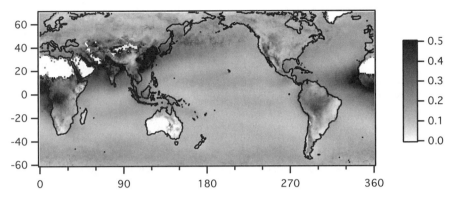

Annual average aerosol optical depth (a measure of the abundance of aerosols) as a function of latitude and longitude, for the years 2004–2008 (measurements were made by the Moderate Resolution Imaging Spectroradiometer onboard NASA's Aqua satellite and were obtained from the NASA Goddard Earth Sciences Data and Information Services Center). White areas are regions where no data were obtained. (See Color Plate 6.7.)

biomass burning over central Africa, and an aerosol soup that originates over Asia and is blown eastward over the Pacific toward the United States.

This means that, although the global average radiative forcing from aerosols is -1.2 W/m^2, this is not evenly distributed over the globe. In regions where aerosol abundance is high, aerosols can have a local radiative forcing of many times this value, whereas in regions that have no aerosols, the local radiative forcing can be zero. This can be contrasted to greenhouse gases such as carbon dioxide or methane, which are well mixed in the atmosphere because of their long atmospheric lifetimes (many years). As a result, their radiative forcing is evenly distributed across the globe.

From a climate perspective, it is important to note that the negative radiative forcing that is due to aerosols offsets 40% or so of the positive radiative forcing that is due to greenhouse gases. In this way, aerosols benefit us, because without them the net radiative forcing would be much higher – and global warming would be worse. But aerosols are not all good – they are also one of the main components of "air pollution" and are a significant health problem for people forced to breathe them. As Figure 6.7 shows, the United States and Western Europe have low levels of aerosol abundances despite large economies because of aggressive air-pollution control efforts. It is the poorer countries with lax environmental regulations, such as China and India, which are responsible for much of the aerosol burden in the atmosphere.

As these poorer countries begin to clean up their atmosphere, we can expect to see the amount of aerosols in the atmosphere diminish. Although such reductions improve public health, they also make global warming worse by reducing the cooling that aerosols provide. This is an unfortunate example of how improvements in one aspect of the environment can lead to problems elsewhere.

6.2.3 Land-use changes

Humans have been remaking the face of the planet for thousands of years, and over the past few centuries, humans have altered vast swaths of the surface. For example, in 1750, approximately 7% of the global land area was under cultivation or pasture; in 1990, that number was slightly more than a third. Such alterations in the surface can modify the Earth's climate. Agricultural land typically has a higher albedo than does the natural landscape, especially if the latter is forest. Thus, cutting down a forest and replacing it with grassland for grazing cattle will increase the surface's albedo. Over all, human *land-use changes* have tended to cool the planet, with a radiative forcing of -0.2 W/m^2.

Another way human activities can alter the albedo of the surface is through the release of black carbon – basically soot. In the previous section we explored how these dark particles have a warming effect when suspended in the atmosphere because they absorb sunlight that falls on them. However, their climate impact does not end there. After a month or two in the atmosphere, these particles are removed from the atmosphere. If they are deposited onto a bright surface, like snow, then they will reduce the albedo of the surface, thereby increasing the absorption of solar energy. With the increase in industrial activities over the past two centuries, the amount of black carbon deposited on snow has led to a positive radiative forcing with an estimated magnitude of $+0.1$ W/m^2.

6.2.4 Changes in the Sun

A final radiative forcing – and one that's entirely unrelated to humans – comes from changes in the amount of solar energy reaching the Earth. As we will explore in more detail in Chapter 7, the Sun is not a constant source of energy; it has well-documented variations. The best known is the 11-year sunspot cycle, over which the solar constant varies by roughly 1 W/m^2 (out of 1,360 W/m^2, so it's not a terribly big variation). The Sun's output also varies slightly with a period of 27 days, which is the time it takes the Sun to rotate once on its axis. Because of the large thermal inertia of the climate system (which is responsible for the lag of several decades between a forcing and the final temperature response), the 11-year and 27-day variations in the Sun have little effect on the climate.

However, it's possible that, over longer periods, the Sun's output can vary – and that this might have an important influence on climate. Unfortunately, our knowledge of these longer-term variations is poor. Accurately measuring the output of the Sun requires doing it from orbit, and that means that we don't have reliable measurements of the solar constant before the late 1970s, when the first satellites designed to measure the Sun's output were launched. Scientists have, however, attempted to reconstruct the long-term record of the solar constant by using proxy data, such as sunspot number. This work has suggested that the Sun may have gotten slightly brighter over the past 250 years, particularly early in the 20th century, and provided a positive radiative forcing estimated to be $+0.12$ W/m^2.

Summing all of the radiative forcings, in the period between 1750 and 2005 we get a net radiative forcing of $+1.6$ W/m^2. This total is the sum of positive forcings, primarily the greenhouse gases, and negative forcings from aerosols and land-use changes. It is worth noting that the total net forcing, $+1.6$ W/m^2, is very close to the radiative forcing from carbon dioxide alone, which is $+1.66$ W/m^2. This, however, is just a coincidence.

Let me emphasize what this value means: If we made all of the changes in green-house gases, aerosols, land use, and so on instantaneously and without letting the atmosphere warm, $E_{in} - E_{out}$ would have increased 1.6 W/m². In order for the Earth to once again be in energy balance, the planet would have to warm, which would increase E_{out}. Earlier in the chapter I said that, for the present-day Earth, E_{in} exceeds E_{out} by approximately 0.9 W/m². This means that the planet has warmed up enough in the past 250 years to erase 0.7 W/m² or so of the imposed radiative forcing. Moreover, the planet needs to continue to warm to erase the remaining radiative forcing.

Another important take-away message from this section is that climate change is about much more than carbon dioxide. Carbon dioxide is indeed the most important greenhouse gas that humans are adding to the atmosphere, and the most important climate forcing of the past few centuries. But other greenhouse gases that humans are adding are also making important contributions to the radiative forcing. Non-gas constituents, such as aerosols, also play an important role.

In this section you should be able to see the foundation of the policies we might undertake to stabilize our climate. As should be clear from this chapter, this requires stabilizing net radiative forcing. There are two obvious ways to do this. First, we can stop activities that produce positive radiative forcing – for example, we can stop emitting greenhouse gases to the atmosphere. In the policy world, this is known as mitigation. Alternatively, we could intentionally engage in activities that produce negative radiative forcing – for example, we could intentionally add sulfate aerosols to the atmosphere – thereby canceling out the positive radiative forcing. This latter approach is what is commonly referred to as geoengineering. I will have much more to say about both approaches in Chapters 11 and 12.

6.3 Feedbacks

One of the ultimate goals of climate science is to make quantitative predictions of the future climate. As discussed in Chapter 4, given the solar constant, albedo, and atmospheric composition, it is conceptually easy to calculate the temperature of a planet. You may expect, therefore, that it would be pretty easy to calculate how much warming we should expect if we added some greenhouse gases to the atmosphere or applied some other specified radiative forcing to the planet.

For example, consider a hypothetical planet with no atmosphere ($n = 0$) and an Earth-like solar constant ($S = 1,360$ W/m²) and albedo ($\alpha = 0.3$). The surface temperature of this planet is, according to Equation 4.5, $T = 255.0$ K. Now imagine that greenhouse gases are added to the atmosphere, so that n increases to 0.01. According to Equation 4.5, the new temperature would be 255.6 K, which is a warming of 0.6 K.

That answer would be correct if nothing else in the climate system changed. But as the planet warms from 255.0 K to 255.6 K, other things do change. For example, because ice melts reliably at 0 °C, warming global temperatures reduce the amount of ice on the Earth's surface. If the melting ice uncovers a dark surface, such as ocean, then this decreases the average planetary albedo (i.e., makes the planet less

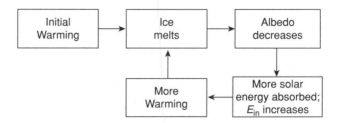

Fig. 6.8   The ice–albedo feedback loop.

reflective), which leads to more absorption of solar radiation and additional warming. This additional warming leads to even more melting, which leads to further decreases in albedo and further warming, and so on. This is known as a *feedback loop*, and it is shown schematically in Figure 6.8.

The net effect of the feedback loop shown in Figure 6.8 is to amplify the initial warming from the addition of greenhouse gas. To provide a quantitative demonstration of this, let's make the reasonable assumption that the albedo is set by the planet's surface temperature, and that the relation between these two variables is $\alpha(T) = 0.3 - (T - 255)/150$. This relation quantifies the decrease in albedo as the planet warms and ice melts.

The equation for surface temperature now includes an albedo that is a function of T:

$$T = \sqrt[4]{\frac{(n+1)\,S\,[1-\alpha\,(T)]}{4\sigma}} = \sqrt[4]{\frac{(n+1)\,S\,\{1-[0.3-(T-255)/150]\}}{4\sigma}} \tag{6.3}$$

This is no longer a trivial equation to solve because T appears on both sides of the equation. Rather, this must be solved by factoring a fourth-order polynomial or by solving the equation numerically.

Returning to our example, we can now calculate the temperature of a planet with $n = 0$ and $S = 1,360$ W/m^2 (we no longer specify albedo because it is a function of temperature). Solving this equation numerically, we can find that the temperature is 255.0 K, the same as when we assumed the albedo was $\alpha = 0.3$. This is not an accident because we defined the albedo to have a value of 0.3 at $T = 255$ K.

Now let's add the same amount of greenhouse gases to the atmosphere as we did in the no-feedback example, which increases n to 0.01. The solution to Equation 6.3 is $T = 256.6$ K, which is a warming of 1.6 K. In other words, the inclusion of the ice–albedo feedback has increased the warming from the addition of greenhouse gases from 0.6 K to 1.6 K. This amplification of the warming is referred to as a *positive feedback*.

You can also have a *negative feedback*. Imagine a planet that is covered with flowers of two colors: white and black. Let's further imagine that, as the temperature of the planet goes up, the white flowers prosper while black flowers die. This means that, as a planet warms, the planet also becomes whiter, that is, the albedo goes up. This increase in albedo will offset some of the initial warming.

To quantitatively explore this, let's assume that the relation between albedo and temperature on this flower-covered planet is $\alpha(T) = 0.3 + (T - 255)/150$, so that an increase in T leads to an increase in α. In this case, an increase in greenhouse gases

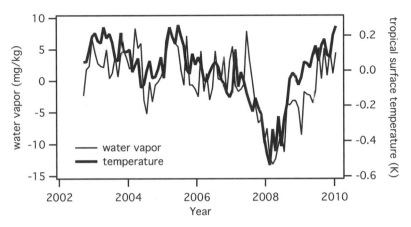

Fig. 6.9 Time series of tropical water vapor and tropical surface temperature. Plotted values are monthly average anomalies, calculated relative to the entire time series and averaged over 30 °N – 30 °S. Water is measured at 300 hPa, which corresponds to an altitude of approximately 10 km (measurements were made by the Atmospheric Infrared Sounder onboard NASA's Aqua satellite and were obtained from the NASA Goddard Earth Sciences Data and Information Services Center).

from $n = 0.0$ to $n = 0.01$ leads to a new surface temperature of 255.4 K, which is a warming of only 0.4 K. This is less warming than we would get if the albedo were constant. Thus, the flower feedback is negative.

6.3.1 Fast feedbacks

There are several important feedbacks in our climate system. The feedback just described above, in which warmer temperatures melt ice, leading to additional warming, is known as the *ice–albedo feedback*. Another important feedback is the *water-vapor feedback*. This feedback arises because a warmer atmosphere can hold more water vapor. Thus, a warming of the surface increases atmospheric humidity, and because water vapor is itself a greenhouse gas, this leads to additional warming. This relation between temperature and atmospheric water vapor is shown in Figure 6.9. Both the ice–albedo and water-vapor feedbacks are positive, meaning that they amplify an initial warming.

There are also negative feedbacks in the climate system. The biggest negative feedback is known as the *lapse-rate feedback*. Because power radiated by a blackbody is equal to σT^4, a warmer atmosphere radiates more power to space. Therefore, as the atmosphere warms, the enhanced radiation offsets some of the initial warming.

The biggest debate among scientists today is about *cloud feedback*. Clouds affect the climate in two opposite ways. First, they reflect sunlight back to space, reducing energy in, which tends to cool the climate. Second, they absorb infrared radiation emitted by the surface, decreasing energy out just like a greenhouse gas, and this tends to warm the climate. The net effect of clouds on the climate is therefore the difference of these two opposing effects.

In our present climate, the reflection of solar radiation is slightly larger than the heat trapping effects, so clouds reduce net E_{in} to the Earth by roughly 20 W/m^2.

This could, however, change in the future. If, in response to an initial warming, the cooling effect of clouds is enhanced, then cloud radiative forcing will become more negative and clouds will act to reduce the initial warming and therefore be a negative feedback. In contrast, if the heat trapping is enhanced, then cloud radiative forcing will become less negative and clouds will amplify the initial warming and therefore be a positive feedback. Although the exact answer is uncertain, the best guess of the scientific community is that the climate feedback is slightly positive, meaning that clouds will slightly amplify an initial warming.

Feedbacks discussed in this section are known as *fast feedbacks* because they occur rapidly in response to a change in surface temperature. Water vapor, clouds, and the lapse rate all respond rapidly to changes in surface temperature, and therefore their impact on energy in and energy out are nearly instantaneous. The response time of ice depends on the type of ice being considered. Some types (e.g., sea ice) will respond in a few years, so changes in these types could also be considered a fast feedback.

6.3.2 Slow feedbacks

In contrast to fast feedbacks, *slow feedbacks* include processes that respond slowly (decades or longer) to increasing surface temperature, so they require a long period of warmth before they significantly alter energy in or energy out. For example, ice sheets are so big that they will likely take centuries or longer to significantly respond to a change in temperature. The ice–albedo feedback associated with ice sheets would therefore be categorized as a slow feedback.

Another slow feedback revolves around the fact that there are large amounts of carbon stored in the ground. One of these carbon reservoirs is permafrost. Permafrost, as the name suggests, is ground that is frozen year-round. Much like that frozen dinner that's been in your freezer since Clinton was President, dead organic plant matter frozen into the permafrost does not decay; it is kept intact as long as the ground remains frozen. If a warming climate leads to the melting of permafrost, then the organic matter in it thaws out and decays, releasing the carbon back into the atmosphere in the form of either methane or carbon dioxide.

We know that permafrost is indeed melting, which is consistent with the large warmings observed in the Arctic over the past few decades (see Chapter 2, Figure 2.2b). In Alaska, for example, roads and buildings constructed on permafrost under the assumption that the permafrost would never melt are now suffering damage as the permafrost melts and the ground underneath begins shifting. In Siberia, permafrost formed during the last ice age is also melting, revealing frozen oddities such as intact wooly mammoths that died 20,000 years ago.

Another source of frozen greenhouse gases is what are known as methane clathrates – methane molecules that are embedded in ice. Clathrates can exist on land or under the ocean, and as with permafrost, warming temperatures can melt the ice and release the methane trapped therein. Given that a molecule of methane is 20 times as powerful a greenhouse gas as a molecule of carbon dioxide, the rapid release of significant amounts of methane is a worrying possibility. And there are several other reservoirs of carbon, such as the tropical forests, which may release the carbon to the atmosphere as the climate warms.

This opens the possibility of a *carbon-cycle feedback*, in which an initial warming leads to the release of large amounts of carbon dioxide and methane that are currently frozen into the ground or otherwise sequestered. The release of these greenhouse gases leads to more warming, and the further release of greenhouse gases. The occurrence of such a feedback in the next century is speculative, but there is reasonably strong evidence that they have occurred in the past, such as during ice-age cycles (which I will discuss in Chapter 7).

Another slow feedback involves vegetation. It has long been known that the distribution of vegetation on the Earth's surface is governed to a large extent by the climate – through the distribution of precipitation, temperature, sunlight, and other such factors. Recently, however, it has been realized that changes in vegetation can also affect the climate. For example, the conversion of a forest to grassland will increase the albedo (because the forest is darker than the grassland), thereby tending to cool the climate. Changes in vegetation can also directly impact exchanges of heat, water, and momentum between the surface and atmosphere, or modify the rate of uptake of carbon dioxide by the vegetation. This introduces the possibility of *vegetation feedbacks* in which changes in climate lead to changes in vegetation, which in turn lead to additional changes in climate.

Probably the slowest feedback is the weathering thermostat. As the Earth's surface warms, the total amount of rainfall will also increase. The increase in rainfall in turn increases the rate of chemical weathering, which removes carbon dioxide from the atmosphere (we explored this in Chapter 5). The reduction in atmospheric carbon dioxide acts to offset some of the initial warming, so this is a negative feedback that tends to stabilize the Earth's temperature. That's the good news. The bad news is that the weathering thermostat operates on geologic time scales, so it only has an impact on the climate over millions of years. We should not expect the weathering thermostat to ameliorate warming over the next century – or over any time scale that we care about.

In general, slow feedbacks are much more uncertain than fast feedbacks because they are so slow that modern Earth science, which is only a few decades old, simply does not have data extending over centuries or millennia needed to observe and understand them. Thus, these slow feedbacks should be considered speculative, and the net effect of slow feedbacks on the climate to be uncertain. Nevertheless, they continue to compel our attention because many of the worst-case climate scenarios involve slow feedbacks causing extremely large warming over the coming century and beyond.

An aside: Feedback vs. forcing

It is worth explicitly discussing the differences between climate feedbacks and radiative forcings. Feedbacks are processes that respond to changes in the Earth's surface temperature, so feedbacks do not initiate climate change. Rather, positive feedbacks amplify and negative feedbacks ameliorate an initial warming. Water vapor, for example, is considered a feedback because the amount of water vapor in the atmosphere is set by the surface temperature of the Earth. If the surface temperature increases, then the amount of water in the atmosphere will also increase, leading to additional warming.

Radiative forcings, in contrast, affect the climate but are themselves unaffected by the climate. The changes in carbon dioxide, methane, and the like between 1750 and 2005 are fundamentally unrelated to the Earth's temperature; instead they are driven by economic growth of human society. Ditto for aerosols.

A confusion arises because some things can be both feedbacks and forcings. For example, although carbon dioxide has been a forcing over the past two centuries, it can also be a feedback if warming temperatures lead to the release of carbon dioxide, as likely happens during ice-age cycles (to be discussed in Chapter 7). Changes in vegetation are a forcing when humans are modifying the vegetation, and they are a feedback when it is the climate that causes the modification.

In some cases it is not clear whether a change is a feedback or a forcing. For example, the processes that regulate stratospheric water vapor are not well understood. As a result, we do not know if changes in stratospheric water vapor that are now occurring are due directly to human activities, so they would be a forcing, or are tied to changing surface temperature, so they would be a feedback. In this chapter, we put it in the forcing category, but more subsequent scientific research may reveal that it belongs in the feedback category. Or perhaps it belongs in both categories.

6.4 Climate sensitivity

6.4.1 Feedback math

Before we define the *climate sensitivity*, I will first go over some basic feedback math. Consider a feedback loop, such as the ice–albedo feedback (shown earlier in Figure 6.8). We will express the strength of this feedback as g, which is the additional fractional warming produced by one trip through the feedback loop per degree of initial warming. Thus, in response to an initial warming ΔT_i, the first trip through the feedback loop produces additional warming of $g\Delta T_i$. However, the feedback also operates on this additional warming $g\Delta T_i$, and this produces an additional warming of $g(g\Delta T_i) = g^2\Delta T_i$. And feedbacks operate on this additional warming too, leading to an additional warming of $g^3\Delta T_i$, and so on. This goes on forever, so the final warming ΔT_f is

$$\Delta T_f = \Delta T_i + g\Delta T_i + g^2\Delta T_i + g^3\Delta T_i + g^4\Delta T_i + \cdots \tag{6.4}$$

We can write this more compactly as

$$\Delta T_f = \sum_{k=0}^{\infty} g^k \Delta T_i \tag{6.5}$$

This infinite series can be rewritten more simply as

$$\Delta T_f = \frac{\Delta T_i}{(1 - g)} \tag{6.6}$$

If $g = 0$, then there is no feedback and the final temperature change is equal to the initial temperature change. If g is between 0 and 1, then ΔT_f is larger than ΔT_i, meaning the feedback is positive. If g is less than 0, then ΔT_f is less than ΔT_i, meaning the feedback is negative.

As the positive feedbacks get stronger and g approaches 1, the denominator in Equation 6.6 approaches 0 and ΔT_f approaches infinity. This should make sense from visual inspection of Equation 6.4: If $g = 1$, then each subsequent term is as big as the previous one and, because the series is infinite, the sum must also be infinite. This situation is sometimes referred to as a "runaway greenhouse effect." In other words, an initial temperature perturbation leads to an infinite temperature rise.

It turns out that the total feedback parameter g for our climate can be expressed as a sum of feedback parameters from the individual feedbacks:

$$g = g_{ia} + g_{wv} + g_{cloud} + g_{lr} \tag{6.7}$$

where g_{ia} is the ice–albedo feedback, g_{wv} is the water-vapor feedback, g_{cloud} is the cloud feedback, and g_{lr} is the lapse-rate feedback (we consider here only the fast feedbacks).

The strongest feedback is the water-vapor feedback, with a magnitude $g_{wv} = 0.6$. This feedback is big enough that, by itself, it would more than double the initial warming ΔT_i. The ice–albedo feedback is substantially weaker, with a magnitude $g_{ia} = 0.1$. Because it is a negative feedback, the lapse-rate feedback has a negative magnitude $g_{lr} = -0.3$. Finally, the cloud feedback is quite uncertain, but most scientists would put its magnitude $g_{cloud} = 0.0–0.3$.

Summing these individual feedbacks, we get a total feedback parameter for our climate of $g = 0.4–0.7$. This means that, for our climate, $\Delta T_f = 1.6–3.3 \, \Delta T_i$. Thus, feedbacks substantially amplify the warming from that due directly to greenhouse gases and other radiative forcings.

6.4.2 Sensitivity

For historical reasons, climate sensitivity is almost always expressed as the warming that occurs if the carbon dioxide is instantaneously increased from 280 ppm, the pre-industrial value, to 560 ppm, twice the pre-industrial value, and then one lets the climate reach a new equilibrium, which takes a century or two.

For this doubling of carbon dioxide, the initial warming ΔT_i is 1.2 °C. Using the feedback strengths from the previous section implies a range of final temperature $\Delta T_f = 2–4$ °C. A more sophisticated analysis by the IPCC concludes that the climate sensitivity is likely in the range 2 °C to 4.5 °C, with a best estimate of approximately 3 °C. The panel also concludes that the sensitivity is very unlikely to be less than 1.5 °C. Values substantially higher than 4.5 °C cannot be excluded, but the evidence supporting such high climate sensitivities is weak.

Doubled carbon dioxide corresponds to a radiative forcing of roughly 4 W/m². Given this, we can also express the sensitivity as the warming per unit of radiative forcing. The climate sensitivity in these units is 0.5–1.1 °C/(W/m²), with a best estimate of 0.75 °C/(W/m²). As an example, in Section 6.2 we calculated that the radiative forcing for a 5% increase in solar constant is $+12$ W/m². Given that, we can calculate how much warming we get by multiplying this radiative forcing by the best-guess climate sensitivity of $0.75\,°C/(W/m^2)$, which yields warming of $9\,°C$.

So far, we have only considered fast feedbacks in our calculation of the climate sensitivity. This is probably appropriate for climate change over the next century. Over longer time periods, though, such as the next millennium and beyond, the contribution of slow feedbacks can become important. These feedbacks are thought to be mainly positive, so the climate sensitivity may be significantly higher when we consider such longer periods. Exactly how much higher is unknown, but looking at previous long-term warming events in response to greenhouse gas emissions (such as the Paleocene–Eocene Thermal Maximum, or PETM, mentioned in Chapter 2) provides some evidence that slow feedbacks may as much as double the climate sensitivity. That would be very bad news for our descendants living in the second half of this millennium.

6.5 Chapter summary

- A radiative forcing is an imposed change on the energy balance of the Earth; in response, the Earth changes its temperature so that energy balance is reestablished. It is important to remember that the radiative forcing is the change in energy balance for the planet (energy in minus energy out) after the imposition of the specific change in the climate, but before the climate has changed in response.
- Because of the Earth's thermal mass, the Earth's temperature takes several decades to respond to an imposed radiative forcing (e.g., increase in greenhouse gases, change in the solar constant).
- The increase in greenhouse gases since 1750 has imposed a radiative forcing of $+3.0$ W/m^2. The increase in carbon dioxide is responsible for $+1.66$ W/m^2, or 55% of the total. The change in aerosols since 1750 has imposed a net radiative forcing of -1.2 W/m^2. This means that aerosols offset approximately 40% of the radiative forcing from increasing greenhouse gases. Summing all changes, we get a net radiative forcing over this time period of $+1.6$ W/m^2.
- Positive feedbacks amplify and negative feedbacks ameliorate an initial warming. For the problem of modern climate change, we are mainly concerned with the following fast feedbacks: water vapor, ice–albedo, lapse rate, and clouds. Together, they double to triple an initial warming.
- Feedbacks are processes that respond to changes in the surface temperature, whereas forcings are unrelated to the surface temperature. Thus, feedbacks do not initiate climate change, but forcings do.
- The Earth's climate sensitivity, which is conventionally defined as the equilibrium temperature increase caused by a doubling of carbon dioxide from 280 ppm to 560 ppm, is 2.0–4.5 °C, with a best-guess value of 3.0 °C. In terms of radiative forcing, the climate sensitivity is 0.5–1.1 °C/(W/m^2), with a best estimate of $0.75\,°C/(W/m^2)$.

Terms

Carbon-cycle feedback
Climate sensitivity
Cloud-condensation nuclei
Cloud feedback
Committed warming
Direct radiative effect of aerosols
Fast feedbacks
Feedback
Ice–albedo feedback
Indirect effect of aerosols
Land-use changes
Lapse-rate feedback
Radiative forcing
Ship tracks
Slow feedbacks
Thermal inertia
Vegetation feedback
Water-vapor feedback

Problems

1. List the important fast feedbacks operating in our climate. Identify whether each is positive or negative.
2. Define climate sensitivity. What is the currently accepted value for our climate?
3. Imagine that we add some carbon dioxide to the atmosphere and the Earth warms by 1 °C. How much warming would there have been if there were no feedbacks?
4. Imagine that our Sun brightens by 1% instantaneously.
   a) How long would it take for the Earth to reach its new equilibrium temperature? Is this longer or shorter than the time it would take Mars or Mercury to reach their respective equilibrium temperatures?
   b) What radiative forcing does this change correspond to?
   c) Approximately how much warming would this brightening cause eventually cause?
   d) How would the calculated radiative forcing change if the brightening takes place over 1,000 years instead of instantaneously?
5. Explain why water-vapor changes are considered a feedback and not a forcing.
6. The albedo changes from 0.3 to 0.31 on the Earth. What is the radiative forcing associated with this change?
7. If doubled carbon dioxide has a radiative forcing of 4 W/m^2, how much of a change in albedo is required to completely cancel a doubling of carbon dioxide on the

Earth (put another way, how much of a change in albedo is required to generate a radiative forcing of -4 W/m^2)?

8. Imagine that, in addition to the fast feedbacks discussed herein, there was a fast negative "flower" feedback like that described in this chapter (as the planet warms, white flowers prosper while black ones died out), and that it had a magnitude $g = -0.3$. What would the Earth's climate sensitivity be in that case?

9. Assume that the Earth has warmed by $5\,°$C since the last ice age, and the change in radiative forcing over that time was $+6.7$ W/m^2. On this basis, calculate the climate sensitivity.

   a) Express the climate sensitivity in $°$ C/(W/m^2).

   b) Express the climate sensitivity in degrees Celsius per doubled CO_2.

7

Why is the climate changing?

In Chapter 2, we detailed the overwhelming evidence that the Earth's climate is changing – evidence so overwhelming, in fact, that virtually no one disputes this anymore. Instead, much of the most heated argument is over the cause of the warming: Is it caused by human activity, or is it natural? In this chapter, we address this question.

Attributing the cause of a trend is more difficult than identifying the trend. Our strategy here is to examine the mechanisms that have changed climate in the past and examine each of them to determine if they could be the cause of the recent warming. You will see that a careful review of all of the possible causes of the recent warming yields the conclusion that the most likely explanation is the increase in greenhouse gases in our atmosphere, which we learned in Chapter 5 is due to human activity.

7.1 The first suspect: Movement of the continents

As you probably know, the Earth's continents are moving. Not fast, mind you – they move at about the same rate that your fingernails grow – but over tens of millions of years, this movement, also referred to as *tectonic motion*, can substantially alter the arrangement of the continents across the Earth's surface. Such changes can directly lead to large changes in the climate through several mechanisms.

For example, the location of continents determines whether ice sheets form. The most important requirement for growth of an ice sheet is summer temperatures cool enough that snow that falls during the winter does not melt during the following summer. Land at high latitudes is the most favorable location for this, as evidenced by the fact that the Greenland and Antarctic ice sheets are both located at high latitudes.

Ice sheets matter to the climate because ice reflects sunlight, so the formation of an ice sheet increases planetary albedo, thereby increasing the reflection of solar radiation back to space and cooling the planet. Through the same chain of logic, the loss of an ice sheet will warm the climate. For example, if Antarctica moved toward the equator over the next 100 million years, loss of the Antarctic ice sheet would be likely, thus leading to important climate impacts.

In addition, the location of the continents determines the ocean circulation. The oceans carry huge amounts of heat from the tropics to the high latitudes, so changing the ocean circulation can therefore alter the relative temperatures of the tropics and polar regions. A good example of this happened 30 million years ago when the Antarctic Peninsula separated from the southern tip of South America, thereby opening the Drake Passage. The opening of this passage allowed winds and water to flow unhindered in a continuous path around Antarctica. This intense flow reduced the transport of warm water and air from the tropics to the South Polar region, causing a dramatic cooling of the Antarctic and contributing to the formation of the Antarctic ice sheet.

The movement of the continents can also indirectly affect the climate by regulating atmospheric carbon dioxide. As I discussed in Chapter 5, carbon dioxide is slowly removed from the atmosphere by chemical weathering, which occurs when atmospheric carbon dioxide dissolves in rainwater and reacts with sedimentary rocks. This process both breaks the rocks down and removes the carbon from the atmosphere.

Movement of the continents can change the pattern of rainfall and expose new rock to the atmosphere, changing the locations and rate of chemical weathering – and therefore the amount of carbon dioxide in the atmosphere. For example, 40 million years ago the Indian subcontinent collided with the Asian continent, forming the Himalayas and the adjacent Tibetan Plateau. Changing wind patterns brought heavy rainfall onto the vast expanse of newly exposed rock in these features, and the resultant chemical weathering drew down atmospheric carbon dioxide over a period of tens of millions of years.

Thus, movement of the continents can indeed change the climate – but could this be responsible for the rapid warming of the past few decades? The answer is no. It takes millions of years for continental movement to cause significant climate change. Continental movements cannot significantly modify the climate over decades or centuries.

7.2 The Sun

As we explored in Chapter 4, one of the factors that controls our climate is the solar constant S. If the Sun brightens, then we expect the climate to warm. We might therefore wonder if the recent warming of the climate can be explained by an increase in the output of the Sun.

This is a reasonable question because it is well known that the Sun's output varies on many time scales. For example, solar physicists believe that, over the Sun's 5-billion-year life, as the Sun burned hydrogen and produced helium, the rate of fusion in the Sun has increased as the buildup of helium leads to increases in the density of the Sun's core. This has caused the Sun to become about 30% brighter over this time.

It turns out that we have the data to test whether the Sun has increased its output over the past few decades. Since the late 1970s, instruments on satellites have been measuring the solar constant; their measurements are plotted in Figure 7.1. Over this period, the most significant observed variation is an 11-year cycle, by which total solar energy output varies approximately 0.1 percent.

Because of the enormous thermal inertia of the oceans, the climate does not respond much to these 11-year variations. To understand why, imagine putting a pot of water on the stove and then turning the burner beneath it on and off each second. If you measured the temperature of the water, you'd find that it was not varying each second.

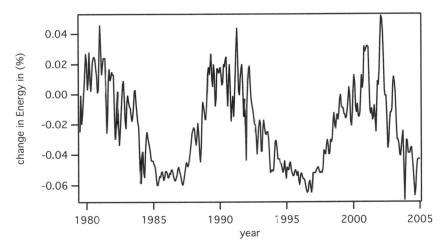

Fig. 7.1   Percentage change in monthly values of energy in (E_{in}), based on the analysis of Frölich and Lean. Seasonal changes in the Earth–Sun distance have been removed (adapted from Forster et al., 2007, Fig. 2.16).

Rather, the temperature of the water would be much more constant. This is why the climate does not respond to 11-year solar variations.

In order for the Sun to be responsible for the recent warming, there would need to be a long-term increase in the solar constant over the past few decades. The measurements show no evidence of this.

Another reason to discount the Sun as an explanation is that an increase in solar output would warm the entire atmosphere. This is not happening. Rather, measurements from weather balloons and satellites show that the stratosphere (the region of the atmosphere beginning at an altitude of 10 km or so) has cooled over the past few decades. Thus, we can conclude with high confidence that the rapid warming of the past few decades is not caused by a brightening of the Sun.

The Sun's influence on climate before the 1970s, however, is more difficult to determine. Solar output for this period must be inferred indirectly from other measurements, such as the number of sunspots, which people have counted for many hundreds of years, or from chemical proxies such as the amount of carbon-14 in plant material. The most recent analyses of these records suggest that the Sun has brightened over the past few hundred years, and this can potentially explain at least some of the gradual warming of the 18th, 19th, and early 20th centuries. As I discussed in Chapter 6, this has led to a positive radiative forcing with an estimated magnitude of +0.12 W/m^2, which is minor compared to radiative forcing from greenhouse gases.

7.3 The Earth's orbit

The solar constant is determined not just by the energy emitted by the Sun, but also by the Earth–Sun distance. If, for example, the Earth moved closer to the Sun, then the solar constant would increase even if the brightness of the Sun did not change.

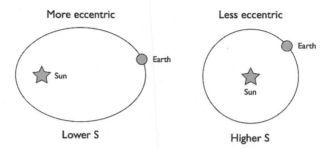

Fig. 7.2 Schematic illustrating how the eccentricity of the Earth's orbit (how elliptical it is) varies with a period of 100,000 years or so.

This is relevant because the Earth's orbit is not a perfect circle: It is an ellipse whose *eccentricity* – the ratio of the length of the ellipse to the width – varies with time.

Over the course of 100,000 years or so, the orbit cycles between an orbit that is slightly more eccentric and one that is slightly less so (Figure 7.2).[1] As the orbit becomes more eccentric, the average Earth–Sun distance increases and the average amount of solar energy falling on the Earth decreases. For the Earth's orbit, the change in eccentricity causes the annual average solar constant to vary by approximately 0.5 W/m^2. This change in the solar constant will lead to changes in the Earth's climate.

Other aspects of the Earth's orbit can also vary, such as the timing of the closest approach of the Earth to the Sun (also known as the *perihelion*). Today, the Earth is closest to the Sun during January, when it is wintertime in the northern hemisphere. Over the next 23,000 years, the date of closest approach will cycle through the entire year. In roughly 11,500 years, the Earth will be closest during July, and in 23,000 years it will again be January. Another important variation is the tilt of the Earth (also known as the *obliquity*). Today, the Earth's spin axis is tilted 23.5 ° from vertical (Figure 7.3). However, over the next 41,000 years, the Earth's tilt will complete a cycle through a range of tilt angles from 22.3 ° to 24.5 °.

Changing the date of closest approach to the Sun or the tilt of the Earth does not change the Earth–Sun distance, so it does not change the solar constant. Rather, these changes change how sunlight is distributed over the planet, in both latitude and season. For example, increasing the tilt of the planet increases the amount of sunlight hitting the polar regions and decreases the amount hitting the tropics. Such changes can alter the climate.

We see in the paleoclimate record nearly perfect agreement between the ice-age cycles over the past few million years (explored in Chapter 2) and the variations in the Earth's orbit. As I discussed earlier in this chapter, the growth of big, continental-scale ice sheets, such as existed during the last ice age, is determined by high-latitude summertime temperatures – because this determines whether snow that falls during

---

[1] When we specify the length of a cycle (e.g., the 100,000-year eccentricity cycle), this is the length for the eccentricity to execute one complete cycle. This means that it takes 50,000 years for the eccentricity to vary from its maximum value to its minimum, and another 50,000 years to return to its starting value.

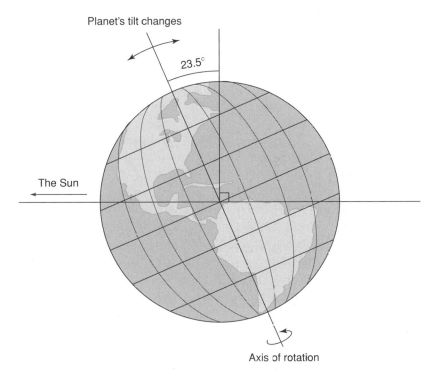

Fig. 7.3   Schematic illustrating how the obliquity of the Earth (the tilt of the spin axis away from a line perpendicular to the orbital plane, the plane defined by the Earth's orbit) varies with a period of 41,000 years.

the winter survives the subsequent summer. Orbital variations regulate how sunlight is distributed over the planet, so they play a role in regulating these temperatures. These orbital variations and the climate effects that follow are often referred to as *Milankovitch cycles*, after Serbian mathematician Milutin Milankovitch, who was the first one to recognize that the ice-ages cycles occurred at exactly the same frequency as variations in the Earth's orbit.

But while these orbital variations are critical in ice ages, are they responsible for the warming of the past few decades? They are almost certainly not. These orbital variations are so slow that it takes at least thousands of years to make any significant change in the amount of or distribution of incoming sunlight. The warming of the past century has been much too fast to be caused by these slow orbital variations. The warming must be due to other causes.

7.4 Internal variability

Changes in the output of the Sun or in the Earth's orbit are examples of *forced variability*, changes in the Earth's climate in response to an imposed change in the planet's energy balance. However, the Earth's climate system is so complex that it can also change without external factors driving it. Such changes, which are driven by the internal physics of the system rather than external changes in the planet's energy in or energy out, are often referred to as *internal variability*.

A good example of internal variability that you can put on your desk is the "drinking bird," a toy in which a toy bird oscillates between standing straight up and rotating over and sticking its beak into a glass of water.[2] The drinking bird relies on internal physics to drive this oscillation; it requires no external forcing.

The best-known example of internal variability in our climate is the El Niño/ Southern Oscillation (ENSO). El Niño events, which make up the warm phase of ENSO, occur every few years and last a year or so. During these events, the Earth warms several tenths of a degree Celsius. El Niño's opposite is La Niña, and during these events the Earth cools several tenths of a degree. These ENSO events cause a temporary temperature change every few years, but no long-term changes in the climate. Figure 2.3, for example, shows the dramatic warming that the Earth experienced during the El Niño of 1998 as well as the cooling experienced during the La Niña of 2008. In fact, many of the short-term variations in the temperature record can be traced back to ENSO events.

ENSO is the dominant and best known internal variation in the climate system. However, other modes of variability are also thought to exist, although they are less well understood. So the relevant question is this: Could the warming of the past few decades be due to internal variability? It is certain that ENSO is not to blame because the entire cycle lasts a few years at most, so it cannot cause a warming trend lasting several decades. Could there be another type of internal variability that might be responsible – one like ENSO, but occurring over decades or centuries instead of years?

The proxy data on climate variation before the past two centuries can help answer this question. Human activities probably had minimal impact on climate before 1800, so climate paleoproxy data before that time should provide a good picture of recent natural climate variability. As we could see in Figure 2.12, the record between 1000 AD and 1800 AD shows nothing similar to the rate and magnitude of warming of the 20th century. Thus, the paleoproxy data do not support internal variability as a cause of the recent warming.

In addition, we can gain insight into natural climate variability by running climate models without any human greenhouse-gas emissions. In these simulations, climate models exhibit variations in global average temperature from year to year and decade to decade that are similar to those seen in the climate proxy data before about 1800, but they produce nothing resembling the rapid warming of the past century.

Finally, no one has identified any mechanism of internal variability that can explain the warming. Ultimately, we cannot exclude internal variability the same way we can definitively exclude, say, a brightening Sun. However, there is basically no evidence supporting it, either. So internal variability is like a suspect in a criminal investigation who has no alibi, but for whom the police have no evidence linking the suspect to a crime. You would be hard pressed to convict him based only on a lack of an alibi.

---

[2] Watch a video here: http://www.youtube.com/watch?v=Yk71GY02diY.

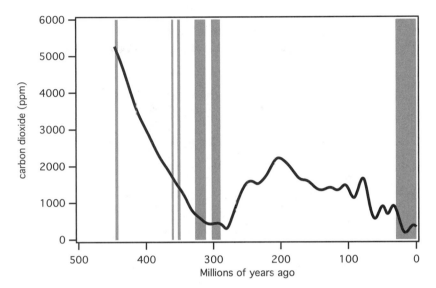

Fig. 7.4   Atmospheric carbon dioxide over the past few hundred million years. The gray bars indicate times when ice existed on the planet (based on Royer, 2006, Figs. 1 and 2).

7.5 Greenhouse gases

Another potential explanation for the recent warming is the increase in greenhouse gases in our atmosphere that has occurred over the past two centuries or so. Chapter 4 discussed the physics of climate and why an increase in greenhouse gases would be expected to warm the planet. This physics is neither new nor complex; it was first recognized in the early 19th century, and our knowledge of it has been refined by nearly two centuries of work by thousands of scientists, including luminaries such as Fourier, Tyndall, and Arrhenius. Moreover, to see the power of greenhouse gases in action, we need only look next door to our neighbor Venus, a planet with a greenhouse-gas-rich atmosphere and a hellishly hot surface temperature to match. Chapter 5 documented the fact that atmospheric carbon dioxide and methane are increasing due to human activities.

In agreement with this simple argument, the geologic record over the past 500 million years shows a strong correlation between temperature and atmospheric carbon dioxide, as shown in Figure 7.4. In particular, widespread ice existed (gray bars in the figure) when carbon dioxide was low, and no ice was found when carbon dioxide was high.[3]

---

[3] Over this same time, the Sun brightened by 6% or so. Thus, the climate associated with a certain level of carbon dioxide a few hundred million years ago would be cooler than the climate would be for the same amount of carbon dioxide today. This explains why glaciations were occurring hundreds of millions of years ago with carbon dioxide abundances higher than today's.

An aside: How does science deals with outliers?

Figure 7.4 also provides a good example of why climate science is hard. Although most glaciations are associated with low carbon dioxide, the eagle-eyed reader will notice that approximately 450 million years ago there was a glaciation when carbon dioxide levels were greater than 5,000 ppm. Such a point is known as an *outlier* – a point that does not agree with scientific expectations. What does this outlier mean for our theory that carbon dioxide drives climate change?

There are several possible explanations for it. First, the theory connecting carbon dioxide with climate may be wrong. Second, the data may be wrong – perhaps there was no glaciation, or maybe carbon dioxide was really much lower than suggested by the proxy data. Third, there may have been something else offsetting the warming from carbon dioxide – for example, massive volcanism could have injected enough sulfur into the atmosphere to lead to low temperatures despite high abundances of carbon dioxide. In this case, because we know so little about what was going on 450 million years ago, we cannot exclude any of these potential explanations.

In his seminal work, *The Structure of Scientific Revolutions*, Thomas Kuhn described how science works. In particular, incorrect scientific theories accumulate anomalies – places where the data do not match observations. These anomalies accumulate until there are so many that the theory is simply no longer tenable, and a scientific revolution takes place in which the old theory is overthrown and replaced by a new one.

It is important to recognize that outliers occur in all fields, not just climate science. For example, you can find – contrary to expectations – people who smoke four packs of cigarettes a day yet who live to be 90 years old. Such anomalies frequently allow scientists to refine and extend their theories: Given that smoking causes cancer, why are some people less susceptible than others? Is it just luck, or is there a physiological basis?

Thus, a single anomaly does not cause scientists to reject the mainstream theory. The question that each scientist must ask individually, and the field must ask collectively, is whether a particular theory has accumulated enough anomalies that it must be replaced by another theory. At present, there are not enough anomalies like the glaciation 450 million years ago to reject the dominant theory that greenhouse gases play a major role in determining our climate. But scientists are always looking for new anomalies – and if they continue to accumulate, eventually this theory of climate will be replaced by another one.

From a practical standpoint, though, no one expects that to occur. The theory that carbon dioxide exerts a strong influence on climate is so successful in predicting so many aspects of our climate that it is quite unlikely that the theory will turn out to be substantially wrong. This is akin to our views on smoking and lung cancer. Although it is possible that future research may disprove the link, it is a very, very unlikely eventuality.

The correlation between greenhouse gases and temperature is particularly clear during an event roughly 55 million years ago known as the Paleocene–Eocene

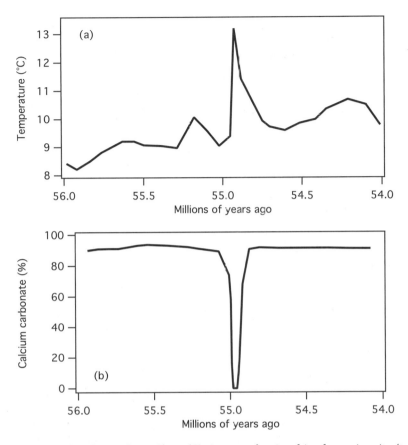

Fig. 7.5 Temperature during the Paleocene-Eocene Thermal Maximum as a function of time from various sites (top panel); calcium carbonate content of ocean sediments (bottom panel) (adapted from Jansen et al., 2007, Fig. 6.2).

Thermal Maximum or PETM, which has been previously discussed in Section 2.2.2. This event began with a massive release of either carbon dioxide or methane. This in turn led to an increase in the Earth's global average temperature of 5–9 °C over the next few thousand years (top panel of Figure 7.5). The mass of carbon was so immense that when the carbon dissolved into the oceans, the oceans became significantly more acidic (as I discussed in Chapter 5, carbon dioxide forms carbonic acid after it dissolves in water). This in turn dissolved calcium carbonate (the material that makes up shells) in the sediments at the bottom of the ocean (bottom panel of Figure 7.5).

The temperatures remained elevated for 100,000 years or so, which is about the length of time it takes the carbon cycle to fully remove the carbon from the atmosphere (I will discuss this in detail in Chapter 8). Interestingly, the amount of carbon released, a few thousand gigatonnes of carbon, is comparable with the amount contained in all of the Earth's fossil fuels. Thus, the PETM is sometimes viewed as a good analog to what will happen if humans burn all of the fossil fuels over the next few centuries. One important difference, however, is that humans are on a pace to release the carbon over several hundred years, whereas it was released during the PETM over several thousand years. Thus, we can expect even more rapid warming than that experienced during the PETM, a period of – geologically speaking – very rapid warming.

The association between carbon dioxide and temperature is even clearer over the past few hundred thousand years. Figure 2.11 shows how carbon dioxide and temperature varied in virtual lock step as the Earth cycled between ice ages and warm interglacials over the past few hundred thousand years. The association between temperature and carbon dioxide, however, is a bit more complicated than the plot may at first suggest. There is strong evidence that ice-age cycles are initiated by small variations in the Earth's orbit (see Section 7.3). However, the changes in sunlight falling on the Earth in response to these slight orbital changes are too small to explain the wide temperature swings during ice-age cycles. Something must be helping the orbital variations produce the observed variations.

What's missing is carbon dioxide. The small initial warming from orbital variations leads to increased levels of carbon dioxide through a mechanism that is not currently well understood. The increase in carbon dioxide leads to further warming. In other words, the orbital variations are the forcing and carbon dioxide is acting here as a feedback that amplifies the small initial warming from the forcing.

> Skeptics' claim: *During the ice ages, carbon dioxide began rising after the temper-ature. This proves that carbon dioxide responds to temperature, and not the other way around.*

The underlying argument here is that if changes in carbon dioxide lag changes in temperature, then carbon dioxide cannot be having any effect on temperature. To see what is wrong with this, let's pretend that you take out a loan of $100. Over the next year (assuming you did not pay any of the balance down), the interest is, say, $10, so the total amount owed grows to $110. In this scenario, the interest lags the debt – there is no interest until *after* you take on the debt. Thus, you can correctly conclude that it is the debt that causes the interest.

However, this does not mean that the interest is not having any effect on the debt. In fact, it is the interest that is causing the debt to grow with time, which in turn is increasing the interest. What we have here is a feedback loop – debt initiates the interest, but the interest thereafter increases the debt. This is exactly the same situation with ice ages and carbon dioxide. Temperature does indeed initiate the changes in carbon dioxide, but the carbon dioxide subsequently causes the temperature to continue to change.

In the case of the ice ages, we are not exactly sure what process releases carbon to the atmosphere as the climate warms. As we learned in Chapter 5, the two biggest sources of carbon for the atmosphere are the land biosphere and ocean. For both of these reservoirs, there are plausible mechanisms that could explain why warmer tem-peratures would release carbon to the atmosphere, but these hypotheses are tentative and our confidence in our understanding of the details of the carbon cycle's response to climate change is low.

Finally, simulations of the 20th century by climate models that exclude the observed increase in greenhouse gases fail to simulate the increase in temperature over the second half of the 20th century. The model run in Figure 7.6a includes natural forcings – primarily changes in the solar constant and volcanoes – but no human

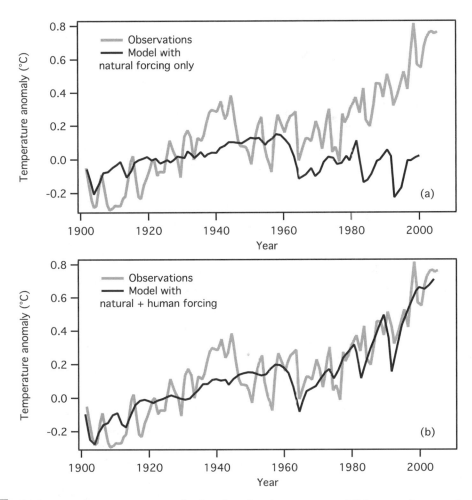

Fig. 7.6 Global mean surface temperature anomalies from the surface thermometer record (lighter curves), compared with a coupled ocean–atmosphere climate model (darker curves). The model includes (a) only nonhuman natural climate forcing, in particular solar and volcanic effects, and (b) natural forcing and human greenhouse-gas emissions, aerosols, and ozone depletion. Anomalies are measured relative to the 1901–1950 mean (the source is Fig. 3.12 of Dessler and Parson, 2010, which is an adaptation of Fig. TS.23 of Solomon et al., 2007).

impact on climate. This calculation reproduces many of the bumps and wiggles in the record, suggesting that these are not due to human activity. But this simulation completely fails to capture the rapid warming that began around 1960.

The model run in Figure 7.6b includes both natural effects as well as the effects of human activities – mainly greenhouse-gas emissions, but also increases in sulfate aerosols and decreases in stratospheric ozone, both of which cool the surface. This model captures the important features in the observations. In particular it captures the rapid warming since 1960 that the model with only natural forcing fails to simulate. This suggests that human greenhouse-gas emissions, volcanic, and solar effects have all contributed to global temperature changes of the past century, but that greenhouse-gas emissions are responsible for most of the rapid late-20th-century warming.

7.6 Putting it all together

As we learned in Chapter 2, the Earth's climate has varied more or less continuously for at least the past several hundred million years, and probably for the entire history of the planet. Obviously, most of these variations have nothing to do with human activities.

Thus, when we consider the recent warming, the first thing we must do is to investigate whether today's warming is due to natural variations. In doing so, most natural explanations can be decisively eliminated (e.g., tectonic motion, orbital variations, variations in the Sun). Internal variability cannot be eliminated, but there is little evidence to support that as an explanation.

In contrast, there is overwhelming evidence supporting the increase in greenhouse gases as the cause of the recent warming. There is strong theoretical evidence that greenhouse gases warm the planet, including the simple arguments detailed in Chapter 4 and the more sophisticated calculations of climate models. There is also observational evidence that carbon dioxide has played a key role in our climate over the past 500 million years.

Taken together with the lack of a competing hypothesis, the case against carbon dioxide's being the cause of the recent warming is strong. Reflecting this, the 2007 report of the Intergovernmental Panel on Climate Change came to the following conclusion: "Most of the observed increase in global average temperatures since the mid-20th century is very likely due to the observed increase in anthropogenic greenhouse gas concentrations." Note that this statement is carefully caveated in three ways.

The first caveat concerns the word *most*. The use of the word *most* makes it clear that greenhouse gases are not the only factor that influences the climate. As we have explored in this chapter, there are other factors that can influence the climate, and some of these (e.g., solar, internal variability) may have been minor contributors to the recent warming. However, increases in greenhouse gases are responsible for more than half of the observed warming.

The second caveat concerns the phrase "since the mid-20th century." It is only during the second half of the 20th century that our observations are sufficient to rule out all of the alternative explanations for the warming. It is certainly possible that greenhouse gases were the dominant cause of the warming during the first half also, but that cannot be proven to the high standards required by the scientific community.

The third caveat concerns the words *very likely*. The Intergovernmental Panel on Climate Change uses a set of carefully defined terms to express confidence. In the parlance of the IPCC, *very likely* denotes a confidence of 90%. This acknowledges that the mainstream scientific view may indeed be wrong (unlike the evidence that the globe is warming, which the IPCC describes as unequivocal). However, the chance is small – about 1 in 10.

For some context, it is useful to look at the previous statements that the IPCC has made on why the climate is warming:

1990: The size of the observed warming "is broadly consistent with predictions of climate models, but it is also of the same magnitude as natural climate variability. Thus

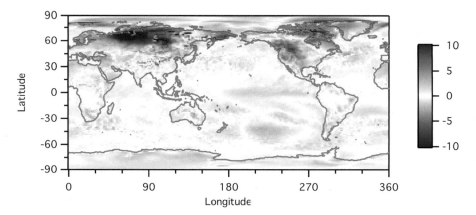

The monthly surface temperature anomaly in December 2009 in degrees Celsius. The reference temperature for the anomaly calculation is calculated at each location as the average of the December temperatures from 2000 to 2009 at that location (data obtained from the ECMWF-interim reanalysis).

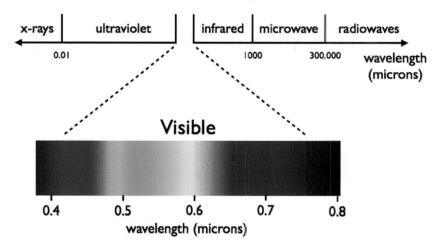

The electromagnetic spectrum. Note that the visible part makes up only a minor part of this spectrum.

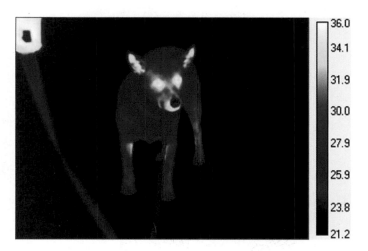

Photo of Bailey the dog in the infrared, with colors assigned to different temperatures. Photo courtesy of New Mexico Tech Department of Physics.

Plate 6.5　Image of a strong temperate cyclone over China, pushing a wall of dust as it moved. The image was captured in early April of 2001 by the Moderate Resolution Imaging Spectroradiometer on NASA's Terra satellite (image obtained from the Earth Observatory; see http://earthobservatory.nasa.gov/IOTD/view.php?id=8341).

Plate 6.6　Ship tracks in clouds off of the West Coast of the United States (image obtained from the Earth Observatory; see http://earthobservatory.nasa.gov/IOTD/view.php?id=37455).

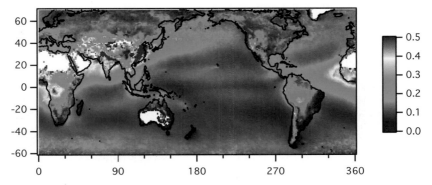

Plate 6.7 Annual average aerosol optical depth (a measure of the abundance of aerosols) as a function of latitude and longitude, for the years 2004–2008 (measurements were made by the Moderate Resolution Imaging Spectroradiometer onboard NASA's Aqua satellite and were obtained from the NASA Goddard Earth Sciences Data and Information Services Center). White areas are regions where no data were obtained.

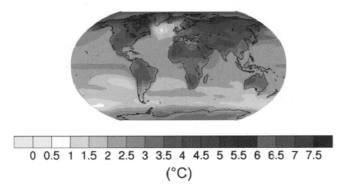

Plate 9.1 Increase in average surface temperature between the end of the 20th century (1980–1999) and the end of the 21st century (2090–2099). This is calculated from an ensemble of climate models driven by the A1B scenario, a mid-level emissions scenario (the source is IPCC, 2007a, Fig. SPM.6).

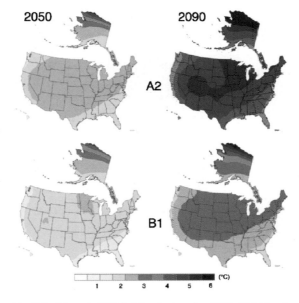

Plate 9.2 Annual warming of the United States in 2050 (left-hand column) and 2090 (right-hand column) relative to the 1961–1979 average for the low-emissions scenario B1 (bottom row) and the high-emissions scenario A2 (top row). These are calculated from an ensemble of climate models (the source is Karl et al., 2009, p. 29).

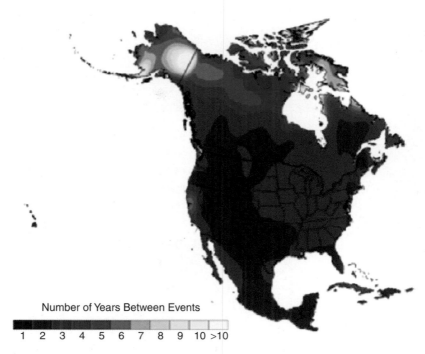

Number of Years Between Events

1 2 3 4 5 6 7 8 9 10 >10

Model predictions, calculated by using the high-emissions (A2) scenario, of the time interval between extremely hot days in 2080–2099 that today occur every 20 years (the source is Karl et al., 2009, p. 33).

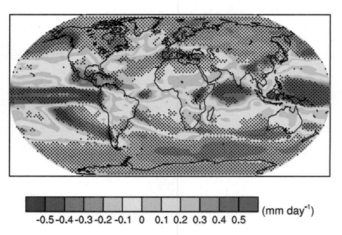

-0.5 -0.4 -0.3 -0.2 -0.1 0 0.1 0.2 0.3 0.4 0.5 (mm day⁻¹)

Change in annual mean precipitation over the 21st century as predicted by climate models driven by a mid-range emissions scenario (A1B). Stippled regions are where at least 80% of models agree on the sign of the mean change (data obtained from Meehl et al., 2007, Fig. 10.12).

the observed increase could be largely due to this natural variability." This statement reflected the fact that climate science was in its infancy at that time. Satellites had only been seriously focusing on measuring climate data for a little more than a decade, computers were slow, and there were few climate scientists in the world working on the problem. As a result, it was not possible at that time to demonstrate a clear human impact on the climate.

1995: "The balance of evidence suggests a discernible human influence on global climate."[4] In the time since the 1990 report, many advances had occurred in climate science. Scientists could now say with confidence that humans were probably having some effect on the climate, although not all of the evidence supported this. The statement did not include any quantification of the human influence.

2001: "[M]ost of the observed warming over the last 50 years is likely to have been due to the increase in greenhouse gas concentrations." This statement is much stronger than the earlier statements, and it reflected improvements in our observations of the climate system, improvements in computers and climate models, and advances in our theoretical understanding of the planet. The 2001 statement is identical to the 2007 statement, except it uses the word *likely*, which denotes a 2 out of 3 chance the statement is true, whereas the 2007 statement uses the words *very likely*, which denotes a 9 out of 10 chance the statement is true.

The evolution of the IPCC reports mirrors the evolution of climate science. And what's most striking, at least to me, is how climate science has actually not changed much. What Arrhenius thought in 1896, what scientists studying the climate thought in the 1950s, 1960s, and 1970s, and what the IPCC report described in 1990 is basically what we think in 2010. The only difference is that our confidence in our understanding of virtually all aspects of climate science has vastly improved.

The overall stability of climate science should provide us with great confidence that it is correct. This is because important scientific ideas are constantly retested by scientists, so the longer an idea survives the more likely it is to be correct. And as a perusal of the IPCC reports shows, most of the major claims of climate science have indeed survived a long time.

It is also worth explicitly stating what is *not* evidence that greenhouse gases are the primary cause of the recent warming. The case for greenhouse gases is not built on the argument that the present temperature of the Earth is exceptional. In fact, we know that the Earth has been much warmer than it is today. As Figure 7.4 shows, over much of the past 500 million years the Earth was so warm that there was no ice anywhere on the planet. Nor is the case for greenhouse gases built on the argument that the rate of today's warming is exceptional. It may be that today's warming is indeed without precedent, but we simply do not have the data covering the entire Earth's history to prove that. Nor is the case built on the occurrence of extreme events, such as the extreme hurricane season of 2005 or the European heat wave of 2003. Rather, the case for greenhouse gases is built on a thorough examination of all of the possible explanations for the recent warming. Such an examination shows that the increase in greenhouse gases is by far the most likely cause.

---

[4] The evolution of this statement is documented in an amusing article by Houghton (2008).

7.7 Chapter summary

- To determine a cause for the present-day warming, we examined all of the natural processes that are capable of changing our climate. Among these were tectonic motions, solar variations, and orbital variations, which can all be decisively rejected as explanations for the present-day warming. Internal variations are harder to reject, but there is no evidence connecting such variations with the present-day warming.

- There is abundant evidence that the increase in greenhouse gases, which is due primarily to human activities, is responsible for the present-day warming. There is strong theoretical evidence that greenhouse gases warm the planet, including the simple arguments detailed in Chapter 4 and the more sophisticated calculations of climate models. There is also observational evidence that carbon dioxide has played a key role in our climate over the past 500 million years. Moreover, detailed calculations made by climate models are only capable of reproducing the warming of the past half-century if the increase in greenhouse gases is included.

- On the basis of this evidence, the IPCC concluded in its 2007 report that "most of the observed increase in global average temperatures since the mid-20th century is very likely due to the observed increase in anthropogenic greenhouse gas concentrations."

Terms

Eccentricity
Forced variability
Internal variability
Milankovitch cycles
Obliquity
Outlier
Perihelion
Tectonic motion

Problems

1. a) List all of the physical processes that can alter the climate.
   b) For all except greenhouse gases, explain why they are unlikely to be the cause of the warming over the past few decades.
   c) List the evidence that greenhouse gases are responsible for the recent warming.
2. What did the IPCC say in its 2007 report about whether humans are causing climate change? What are the three caveats in the statement?
3. Why are feedbacks (e.g., increases in water vapor) not listed as potential causes of climate change?

4. Explain the physical mechanism for the occurrence of ice ages. Make sure you explain the role of carbon dioxide and its timing with respect to the temperature change.

5. Critique this statement: "It is clear that it was warmer around 1000 AD, during the Medieval Warm Period, than it is today. Therefore, humans cannot be causing today's warming." Assuming that the claim that the Medieval Warm Period is warmer than today is correct (it may be, but it's debatable), explain whether this argument is correct.

6. What are the three ways that the Earth's orbit varies? How does each variation affect the climate?

7. Explain how the Paleocene-Eocene Thermal Maximum provides support for the claim that today's warming is caused by humans.

8. How does plate tectonics affect our climate?

The future of our climate

Chapter 4 described the primary factors that control our climate: the composition of our atmosphere (e.g., the amount of greenhouse gases in it), the solar constant, and the albedo. If we know how these factors are changing in the future, we can estimate how the climate will change in response.

Of these factors, the most important changes over the next century are expected to be in the composition of the atmosphere. This means that predicting future climate basically comes down to predicting the amount of greenhouse gas in our atmosphere. This in turn requires that we predict how much greenhouse gas will be emitted into the atmosphere each year from human activities. Such projections of greenhouse-gas emissions, known as *emissions scenarios*, form the backbone of our predictions of climate change over the coming century. In this chapter, I describe how they are constructed and what they tell us about our future climate.

8.1 The factors that control emissions

At its simplest, the amount of greenhouse gas released by a society is determined by the total amount of goods and services consumed by that society. This is true because the production of any good or service – be it a car, an iPod, a university lecture, a cheeseburger, or an hour of computer consulting – requires energy. Furthermore, because most energy is derived from fossil fuels, the production of goods and services therefore leads to the release of carbon dioxide. The emissions of other greenhouse gases also generally scale with the amount of consumption, although the causal linkages may not be as obvious.

The total value of goods and services produced by an economy is known as the *gross domestic product*, generally abbreviated GDP. Thus, total emissions by a society are basically set by that society's GDP. If the GDP doubles, then we expect emissions to double, as long as everything else remains the same. Strong evidence of the link between GDP and emissions can be seen during recessions. During a recession in late 2009, for example, the *New York Times* reported that "Global carbon emissions are expected to post their biggest drop in more than 40 years this year as the global recession froze economic activity and slashed energy use around the world."[1]

Rather than consider GDP as a whole, it is useful to break it into the product of two factors: population and affluence. It should be obvious that GDP should scale

[1] J. Mouawad, "Emissions of CO_2 Set for the Best Drop in 40 Years," *New York Times*, September 21, 2009 (available at http://www.nytimes.com/2009/09/22/science/earth/22emissions.html).

with population. Every person in a society consumes goods and services, so if the population doubles (and everything else remains the same), then total GDP will also double. This also means that emissions must also scale with population – so emissions will double if the population doubles.

In addition to the number of people, how rich each person is also matters because, as people get richer, they consume more. To illustrate the affect of affluence on emissions, consider the following three families. The first family are subsistence farmers living in sub-Saharan Africa. This family lives in a small one-room house without electricity or running water. They do not own a car and are too poor to buy anything but the bare necessities of life. They farm by hand or with a draft animal. Because the members of this family are so poor and consume so little, they are responsible for few greenhouse-gas emissions.

Now consider a family near the bottom of the economic spectrum in the United States. The four members of this family live in an apartment and they share one car. Their apartment is not air conditioned; they own a television and one or two heavy-duty electrical appliances, such as an oven. Compared with the subsistence farming family in Africa, this family is richer and consumes more, and is therefore responsible for more greenhouse-gas emissions.

Finally, consider an upper-class family in the United States. This family lives in a 4,000-ft^2 single-family house and owns three cars (for the husband, wife, and a teenage child). The house has televisions in almost every room, several computers, VCRs, game consoles, and a rich assortment of electrical appliances. Because of the significant consumption allowed by their affluence, this family is responsible for more emissions than the poorer U.S. family and many, many times the emissions of the subsistence farming family.

We need a third factor to convert a level of total consumption, expressed in dollars, to greenhouse-gas emissions. This third term relates how much greenhouse gas is emitted for every dollar of consumption; it is known as the *greenhouse-gas intensity*. Putting these all together, we can now relate emissions to the factors that control it in a simple equation:

$$I = PAT \tag{8.1}$$

Here I is equal to the total emissions of greenhouse gases into the atmosphere (these emissions then cause climate impacts, which is why emissions are represented by the letter I); P is the population, A stands for affluence, and T stands for greenhouse-gas intensity. Affluence A is GDP per person – the average amount of goods and services each person consumes – so the product of P and A is the GDP.

Variations in any of these terms will change the amount of greenhouse gas emitted by a society. For example, as people get richer and their affluence increases, so do their emissions. The decomposition of emissions into these factors is often referred to as the *IPAT relation* or the Kaya Identity.

The greenhouse-gas-intensity term T can be further broken down into two terms:

$$T = \text{EI} \times \text{CI} \tag{8.2}$$

Here EI stands for *energy intensity* – the number of joules of energy it takes to generate 1 dollar of goods and services (CI is defined in later text). Energy Intensity itself is determined by two factors. First is the mix of economic activities in the society. Different economic activities take different amounts of energy to generate 1 dollar of economic output, so it is dependent on the mix of economic activities in an economy. For example, it takes much more energy for a steel mill to produce 1 dollar's worth of steel than for a university to produce 1 dollar's worth of teaching. The steel mill must run blast furnaces and other heavy equipment, whereas the university only requires lighting, air-conditioning, computers, and the like. In other words, industrial manufacturing has a higher energy intensity than do white-collar service-oriented activities.

The second factor in determining the energy intensity of an economy is the efficiency with which the economy uses energy. For any economic activity, there are usually several technologies to accomplish it. For lighting, for example, there is the standard incandescent light bulb (the kind with the filament), the compact fluorescent light bulb, and the newly developed light-emitting diode or LED light bulb. As described in Chapter 3, incandescent light bulbs are dreadfully inefficient, requiring 60 W of power to produce the same light as a compact fluorescent light bulb drawing 14 W. An LED light bulb, in contrast, can produce the same light by using only 2 W of power. These three light-bulb technologies can all light a room, but they consume vastly different amounts of energy doing it. The trade-off is that better technology is often more expensive. LED light bulbs, for example, have higher upfront costs than incandescent bulbs (even though the long-term cost of operation is lower). As a result, it frequently takes a certain level of wealth in order to adopt the most energy-efficient technology.

The other term that determines greenhouse-gas intensity is the *carbon intensity*, or CI, which is the amount of greenhouse gas emitted per joule of energy generated. This term reflects the mix of technologies that a society uses to generate energy, including fossil fuels as well as technologies that do not emit carbon dioxide. Among fossil fuels, combustion of natural gas (methane, or CH_4) produces the least amount of carbon dioxide per joule of energy generated. Thus, it has the lowest carbon intensity, which is one of the reasons it is often considered to be the "greenest" of the fossil fuels. Oil produces more carbon dioxide per joule than methane, so it has a higher carbon intensity. The highest carbon-intensity fossil fuel is coal – it produces roughly twice the carbon dioxide per joule as methane – which explains why many people who are concerned with our climate are opposed to the construction of new coal-fired power plants. Energy sources also exist that produce no carbon dioxide, such as hydroelectric, nuclear, wind, or solar energy sources.

Overall, the carbon intensity of a society reflects the mix of technologies used to generate energy. For a country such as France, which generates much of its electricity from nuclear energy, this term will be smaller than for a country such as China or the United States, which both rely heavily on coal for electricity.

It is useful to go over the units in the IPAT relation to make sure that everything is clearly defined. Population is obviously the number of people. Affluence, which is GDP per person, has units of dollars per person. The product of population and affluence is GDP, which has units of dollars. Energy intensity has units of joules per dollar, and carbon intensity has units of emitted carbon dioxide per joule. Greenhouse-gas intensity is the product of energy intensity and carbon intensity, and therefore it

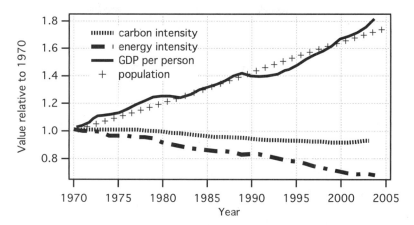

Fig. 8.1 Population, affluence, carbon intensity, and energy intensity for the entire world, relative to values in 1970 (adapted from IPCC, 2007b, Fig. 2).

has units of carbon dioxide emitted per dollar. The product of population, affluence, and technology therefore has units of carbon dioxide emitted.

8.2 How these factors have changed in the recent past

In the previous section, the problem of estimating future emissions of greenhouse gases was deconstructed into the problem of estimating trends in population, affluence, energy intensity, and carbon intensity. Let's look at these terms over the past few decades.

8.2.1 Population

Population has been rapidly increasing for the past few centuries. It took all of human history up to 1804 for the global population to reach 1 billion. The 2-billion-people mark was reached 123 years later, in 1927, and the 3-billion-people marker was reached just 33 years later, in 1960. After that, world population increased by 1 billion people every 12–13 years, reaching 6 billion in 1999 and 7 billion in 2011. Figure 8.1 shows that the population has increased by 80% over the past few decades. Today, world population is increasing by roughly 200,000 people per day, a population growth rate of approximately 1% per year. Most of this growth is occurring in the developing world, where fertility rates remain high.

8.2.2 Affluence

Figure 8.1 shows that affluence, measured as GDP per person, increased by 80% over the past few decades of the 20th century. We can also see discrete political events in the affluence data, such as the 1989 collapse of the Soviet Union and the associated political upheaval in Eastern Europe, as well as various recessions. Moreover, in the past decade, the remarkable economic growth of China (with affluence growing at 10% per year or so) and other large developing countries has played a key role in driving global growth.

8.2.3 Technology

The first part of the technology term, the energy intensity term, has decreased over the past century as our society has developed more efficient ways to use energy (Figure 8.1). Some of this increasing efficiency has been driven by market forces: Because energy costs money, a more energy-efficient piece of equipment or process will reduce costs. Because of this economic pressure, just about everything you buy today is more energy efficient than the comparable 1950s version. Much of this increase in efficiency is incremental, meaning that each new generation of a particular piece of equipment uses slightly less energy than the previous version. Sometimes, however, there is a revolution in technology that greatly reduces energy consumption. A good example is the revolution in lighting technology we are now experiencing. As the world switches from incandescent bulbs to compact fluorescent bulbs and then to LED bulbs, the amount of energy being consumed by lighting will experience a substantial one-time drop.

Changes in the mix of goods and services produced by the world's economy has also led to decreases in energy intensity. Over the past century, the fraction of the world economy based on energy-intensive heavy industry and manufacturing has declined, while the fraction based on services has increased.

Figure 8.1 also plots the carbon intensity, the amount of carbon dioxide released per joule of energy generated. This quantity has decreased slightly over the past few decades as the world shifted from coal to cleaner natural gas. Recently, however, as the world begins to run out of cheap oil and gas, the decline has been arrested as the world switches back to coal.

Thus, increases in population (P) and affluence (A) have tended to increase emissions, while a decrease in greenhouse-gas intensity (T) has tended to decrease emissions. Putting them together, the rate of increase in population and affluence has been much greater than the rate of decrease of greenhouse-gas intensity, so the net change in emissions between 1970 and 2005 was an increase of 75%.

8.3 How these factors will change over the 21st century

8.3.1 Population

Some of the factors that control population are well known. Affluence, for example, strongly determines how many kids a woman has, with the poorest countries having the highest fertility rates. In extremely poor societies, children can be put to work at a young age and are therefore a source of income. This is generally not the case in rich countries, where children are a net drain on family resources until they are adults (trust me on that). In addition, high rates of childhood death in poor countries mean that parents must have many children to ensure that some of them survive into adulthood.

Improvements in health care that occur as a society gets richer, however, mean that rich parents can reasonably expect their children to survive into adulthood. The amount of education that women receive is also a factor, with fertility rates declining as women become better educated and good-paying jobs become available to them as an alternative to child rearing. The factors that control population are not completely understood, however. For example, it has recently been discovered that average family size has started increasing in the richest countries, which was unexpected.

Taking everything we know into account, the most likely scenario is that world population will peak in the second half of the 21st century at between 9 and 10 billion people. However, given our uncertain knowledge about population growth, it is certainly possible that the actual population trajectory could deviate significantly from this.

8.3.2 Affluence

Factors such as the level of education in the population, rule of law, free trade, and access to technology are key factors in determining how fast affluence grows. In general, economic growth rates are highest for countries making a transition out of poverty and into the group of rich countries of the world. For example, economic growth was fastest in the United States in the late 19th century, in Japan after World War II, and in China today. Growth rates are lower for large, advanced societies. Based on these factors, expert predictions are that affluence will increase over the 21st century at 2–3% per year for developing countries and 1–2% per year for industrialized countries.

8.3.3 Technology

Energy intensity has at times decreased as fast as 2% per year, but the periods of fastest decreases occurred during periods of rapid shift in economic mix or as responses to energy price shocks. Over the 20th century, the average decrease was 1% per year for the entire world. Prospects for future progress are likely of a similar magnitude. Continued reductions in carbon intensity are possible through the expanded use of natural gas to replace oil and coal and by shifting new energy supplies to noncarbon sources. However, current energy-market conditions are not supporting this shift, as the recent shift toward coal shows.

8.4 Emissions scenarios

It's hard to make predictions – especially about the future.[2]

Although we have a reasonably accurate idea of the factors that control greenhouse-gas emissions, making accurate predictions of these factors is difficult. For example, predicting future population trends requires predictions of factors such as the rate of

---

[2] This statement has been attributed to various people, including Niels Bohr and Yogi Berra.

poverty, evolution of religious and social views on birth control, the rate of education of women in high-fertility regions, available healthcare in these regions, and so on.

Because it is so difficult to make a single accurate prediction of the factors that control emissions, the community of experts has avoided making a single prediction for the 21st century. Rather, a set of alternative but equally plausible *emissions scenarios* has been created. Each emissions scenario is an internally consistent vision of one way the world might evolve in the future, and the full set of emissions scenarios is designed to span a range of alternative future evolutions of the world. To drive home this point, the scenarios are not referred to as predictions but are instead described as projections. Although this may seem to be an insignificant word change, it is extremely important to keep this distinction in mind when you think about the emissions scenarios.

The most well-known set of scenarios comes from the IPCC. As part of its assessment process in the late 1990s, the IPCC constructed four main emissions scenarios, each based on a different storyline of how the world might evolve over the 21st century. These four families have the extremely unimaginative names A1, A2, B1, and B2.

Summary of scenario families

A1: This storyline describes a future world of rapid economic growth, where both the rich and poor experience gains in wealth, leading to a reduction in the wealth gap between rich and poor and a decrease in poverty. Because of the wealth gains of the poor, population growth diminishes, and global population peaks in mid-century and declines thereafter. There is also a rapid introduction of new and more efficient technologies.

A2: This storyline also describes a world of high economic growth, but the growth is unevenly distributed – with the rich getting richer and the poor remaining poor. Consistent with the poverty of this scenario, the global birthrate remains high, resulting in a population that continuously increases throughout the 21st century. Because of the uneven economic growth, technological development is slower than in other storylines, with deployment of new technologies occurring mainly in the richer parts of the world.

B1: This storyline describes a world where economic growth is evenly distributed between the rich and poor. Economic growth is slower than in the A1 scenario because of an emphasis on sustainable growth and environmental protection. This storyline has the same global population that peaks in mid-century and declines thereafter as in the A1 storyline. The world economy shifts toward a service and information economy, with associated reductions in energy intensity, and the introduction of clean and resource-efficient technologies.

B2: This storyline describes a world where economic growth is unevenly distributed – with the rich getting richer and poor remaining poor. Population increases continuously, although at a rate slower than the A2 scenario, and with less rapid and more diverse technological change than in the A1 and B1

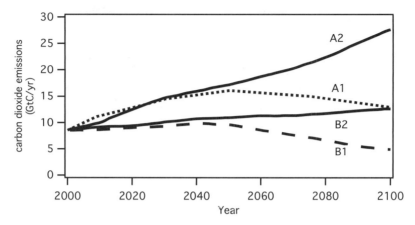

Fig. 8.2 Emissions of carbon dioxide for the four emissions families. The emissions are in GtC per year (adapted from Albritton et al., 2001, Fig. 17).

storylines. Sustainability and environmental protection are the focus at local and regional levels.

To summarize, the "A" storylines (A1, A2) describe worlds with high rates of economic growth. The "B" storylines (B1, B2) describe worlds where economic growth is slower but the Earth's resources are managed in a more sustainable way. The "1" storylines (A1, B1) are worlds in which poverty is reduced and there is an overall convergence between the rich and poor, whereas the "2" storylines (A2, B2) are worlds in which the current split between rich and poor remains.

A key aspect of these storylines is that they are internally consistent, so that the assumptions for population, affluence, and technology all fit together. For example, because people have fewer children as they get richer, the scenarios in which the world's poor become richer (A1, B1) feature slower population growth than the scenarios in which poverty is rampant (A2, B2). Another example is that the development and adoption of new technology requires high economic growth to support it – so the higher the economic growth scenarios (A1, B1) have more rapid adoption of new and cleaner technologies. It should also be noted that the emissions scenarios described by the IPCC all assume that the world makes no explicit effort to address climate change by reducing emissions.

Figure 8.2 shows the rate of emissions of carbon dioxide during the 21st century, in units of GtC/yr. The B1 scenario has the lowest emissions because there is moderate population growth and strong technological development directed toward development and deployment of alternative energy sources that emit little or no carbon dioxide. This scenario predicts that emissions peak in the middle of the 21st century and decline thereafter, reaching 5 GtC/yr or so in 2100, which is approximately half of what humans are now emitting. On the other end of the spectrum is the A2 scenario. That scenario's high population growth, high economic growth directed primarily toward the rich, and slower technological development produces high emissions throughout the century. By the year 2100, emissions for the A2 scenario reach 30 GtC/yr, which is roughly three times what humans are emitting today.

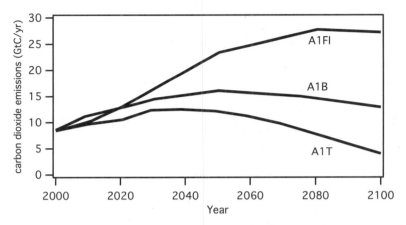

Fig. 8.3   Emissions of carbon dioxide for the three scenarios in the A1 storyline. The emissions are in GtC/yr (adapted from Albritton et al., 2001, Fig. 17).

The A1 family contains two additional scenarios, designated A1T and A1FI, which utilize exactly the same assumptions for population and affluence but sharply different assumptions about technology. The A1T scenario (the letter T stands for technology) assumes a shift toward energy sources that do not emit carbon dioxide over the next few decades. The A1FI scenario (the acronym FI stands for fossil intensive) assumes a shift to coal. These two scenarios are plotted in Figure 8.3, along with the A1B scenario, which assumes a balance between fossil fuels and new technology.

In 2100, the A1FI scenario predicts emissions close to 30 GtC/yr, whereas the A1T scenario predicts emissions close to 5 GtC/yr. The spread between these two scenarios is the same as the spread in emissions between the four scenario families plotted in Figure 8.2. This emphasizes the importance of technology in controlling emissions. It is possible to get an enormous range of emissions simply by changing the assumptions about technology development and implementation.

8.5 Predictions of future atmospheric composition

The next step is to take these emissions scenarios and convert them into atmospheric concentrations of greenhouse gases. This is done by feeding the emissions scenarios into a carbon-cycle model. The carbon-cycle model calculates how much of the carbon dioxide emitted to the atmosphere is absorbed by the ocean and land reservoirs. The remainder stays in the atmosphere and increases atmospheric carbon dioxide.

Figure 8.4 shows the atmospheric carbon dioxide amounts predicted for each scenario. From an abundance of carbon dioxide of 390 ppm in 2010, the IPCC's scenarios project that atmospheric carbon dioxide abundances will be between 550 and 900 ppm in 2100. The lower limit, 550 ppm, from the B1 scenario, represents roughly twice the pre-industrial atmospheric abundance of carbon dioxide. The upper limit, 900 ppm, from the A2 scenario, represents more than a tripling of pre-industrial carbon dioxide abundances.

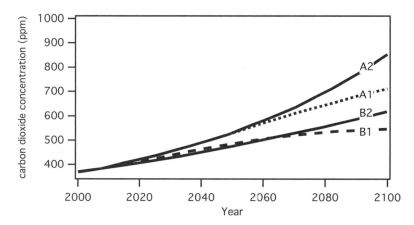

Fig. 8.4 Atmospheric abundances of carbon dioxide (in ppm) for the four emissions families (adapted from Albritton et al., 2001, Fig. 18).

Of course, as discussed in Chapter 6, carbon dioxide is just one of the things that are changing our climate. There are other greenhouse gases (e.g., methane) and non-greenhouse gases (e.g., aerosols, land-use changes) that are also forcing the climate. The same methods used to develop carbon dioxide scenarios are used to develop scenarios for all of these other things that can alter our climate.

Given estimates of how each of these factors will change in the future, we can calculate the radiative forcing from each and then combine them to come up with a total projected radiative forcing. These calculations suggest that, by the end of the 21st century, total radiative forcing would be 4–8 W/m^2 above pre-industrial levels. This is a huge amount of radiative forcing – comparable to the largest variations over the past 500 million years. Such radiative forcings have changed the Earth from an ice-age planet into one that is completely ice free. And carbon dioxide is responsible for 80% of the radiative forcing, which is why it is the single most important factor in future climate change.

8.6 Predictions of future climate

8.6.1 Over the next century

The estimates of atmospheric carbon dioxide abundances shown in Figure 8.4 are then input to climate models, which calculate a future climate for each scenario. Estimates of climates for the B1, A1B, and A2 scenarios, representing low, medium, and high rates of greenhouse-gas emissions, respectively, are shown in Figure 8.5. These model simulations suggest that by the end of the 21st century, the Earth will be 1.8–3.6 °C warmer than the late 20th century.

Also as shown in Figure 8.5 is Line C, which shows the predicted surface temperature if the atmospheric abundance of greenhouse gases and aerosols were stabilized at Year 2000 values. It shows that temperatures will continue to increase by 0.4 °C

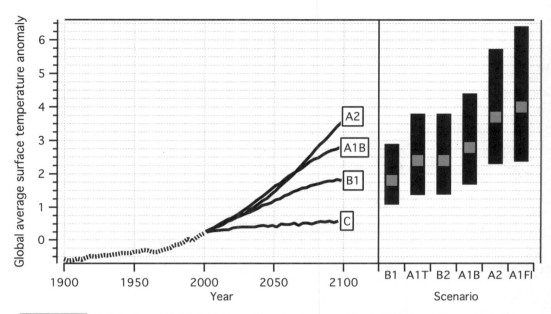

Fig. 8.5 Model estimates of global annual average surface temperature for the B1, A1B, and A2 scenarios (relative to the 1980–1999 average), along with the predicted surface temperature (C) if the atmospheric abundance of greenhouse gases and aerosols were stabilized at Year 2000 values. The bars at right are the likely range of temperatures for each scenario, where the lighter portions are the most likely temperatures. The range of temperatures is derived by using a set of climate models (source is Dessler and Parson, 2010, Fig. 3.14, which was adapted from IPCC, 2007a, Fig. SPM.5).

over the next several decades. This is a significant amount, comparable to the warming of the planet over the past few decades and approximately half the warming over the 20th century. The reason temperatures keep rising even after greenhouse gases stabilize was discussed in Chapter 6: Because of the enormous heat capacity of water, the Earth is not presently in thermal equilibrium and must continue to warm to reestablish energy balance. This warming is essentially unavoidable, which is why it is often referred to as "committed warming" (which is why the line is marked with the letter *C*). And because it cannot be avoided, we must adapt to it. That does not mean that all climate change is unavoidable – much of the warming over the 21st century could still be avoided if we take prompt action.

The right-hand panel in Figure 8.5 shows the likely range of temperatures in 2100 predicted for each scenario, as well as the A1T and A1FI scenarios. This range is generated by taking the same emissions scenarios and running them through a large number of different climate models. Because each model handles the details of the physics of the atmosphere differently, it produces slightly different results.

This allows us to separately estimate the uncertainty in predicted temperature in Year 2100 as a result of uncertainties in emissions scenarios and uncertainty in the physics of the climate models. Figure 8.5 shows that the difference in projected temperature in 2100 between the low-end (B1) and high-end (A2) scenarios is approximately a factor of 2. Figure 8.5 also shows that, for a single emissions scenario, there is a similar factor of 2 difference between the highest temperature predicted by the group of models and the lowest temperature. Thus, the uncertainty in predicted temperature in 2100 is approximately evenly split between uncertainty in emissions and uncertainty in the physics of climate.

Finally, I should say something about the Sun. As we learned in Chapter 7, the output of the Sun can vary, and, if these variations are large enough, they can cause significant climate change. For example, it is generally thought that variations in the output of the Sun played a key role in the Little Ice Age, a cold period that occurred several hundred years ago. The projections in Figure 8.5 all assume no long-term changes for the Sun over the 21st century. If the Sun does become either brighter or dimmer, then that could have an important impact on the evolution of the climate. Unfortunately, we have essentially no ability to predict what the Sun will do, so this remains another uncertainty in out predictions of climate. In the same vein, a large volcanic eruption could also significantly alter the climate. However, such a perturbation would last only a few years, so it is unlikely that a volcano could radically alter the long-term trajectory of the climate.

An aside: Will the evolution of the climate over the 21st century look like the trajectories plotted in Figure 8.5?

Not really. If you look at Figure 2.2 you will see that, over the past 130 years, the climate has generally warmed, but it also shows significant random year-to-year variability, which is caused by things such as El Niño cycles and volcanic eruptions. Such variability means that temperatures can decline for a few years, even as the climate is experiencing a long-term warming. Individual model simulations show this year-to-year variability. The model lines plotted in Figure 8.5, however, are not the result of individual model runs. Instead, each line is the average of many model runs. In the individual model runs going into the average, the highs and lows caused by the short-term variability do not occur at the same time in each model run, so when you average many model runs together, the short-term ups and downs tend to cancel out and you get a smooth increase in temperature throughout the century. In reality, short-term variability is going to be important and we can expect the same kinds of ups and downs seen in the past 130 years to continue to occur in the future.

8.6.2 Climate change beyond 2100

Even though Figures 8.2 through 8.5 stop in Year 2100, we should not take this to mean that emissions and the associated climate change stop in 2100. In fact, many scenarios have significant emissions and warming that extend into the 22nd century and well beyond.

Exactly how long emissions can continue is a fiercely debated point. Everyone agrees that fossil fuels will eventually run out, and emissions from their combustion will therefore cease. Although many experts think that this will happen in the next century or two, the exact timing is disputed, as are estimates of how much carbon can be emitted to the atmosphere. The range of total emissions extends from lower values of 1,500 GtC to more worrying estimates of 5,000 GtC. These estimates are all well above the 300 GtC or so that humans have already emitted into the atmosphere over the past few centuries. And it should be remembered that lags in the climate system, as well as slow feedbacks, have the capacity to continue the warming long past the cessation of emissions.

To get a feel for the long-term evolution of the climate over the next millennium, we must revisit the carbon cycle discussed in Chapter 5. As we learned there, roughly 50% of the carbon dioxide added to the atmosphere is removed within 1 year or so as carbon is absorbed by the land biosphere and the surface water of the ocean. Removing additional carbon dioxide requires transport into the deep ocean, which is a much slower process. After a few centuries, an equilibrium is established between the atmosphere and deep ocean, in which approximately 30% of the carbon emitted from human activities remains in the atmosphere, with the other 70% residing in the deep ocean.

Over much longer time spans, carbon dioxide is slowly removed from the atmosphere–ocean system by chemical weathering. Eventually, after 100,000 years or so, carbon dioxide emitted by humans will be almost entirely removed from the atmosphere. The upshot is that, if we add carbon dioxide to the atmosphere, a signif-icant fraction stays there for a very long time – centuries to millennia and beyond – and the warming it causes therefore also sticks around for a very long time.

This long residence time of carbon dioxide in our atmosphere is shown in Figure 8.6a, which shows atmospheric carbon dioxide over the next 1,000 years for emissions scenarios where atmospheric carbon dioxide increases until it reaches 550, 850, and 1,200 ppm, at which point emissions from human activity decline instantly to zero. Even by Year 3000, eight to nine centuries after carbon dioxide emissions ceased, atmospheric carbon dioxide in all scenarios remains well above pre-industrial values (280 ppm). This is simply a reflection of how long it takes for an addition to atmospheric carbon dioxide to be removed from the atmosphere.

The temperatures associated with each carbon dioxide time series are shown in Figure 8.6b. Even after emissions cease, the temperatures do not decline significantly over the next 1,000 years. This is a consequence of three factors. First, carbon dioxide remains elevated throughout the millennium, so it continues to heat the planet even after emissions stop. Second, the ocean's large heat capacity means that the planet cools off very slowly. This is the flip side of the situation in which the warming lags the carbon dioxide – the cooling will lag any decrease in atmospheric carbon dioxide abundance. Third, the slow feedbacks, such as the loss of the big ice sheets, will act to oppose any cooling.

The important point here is that emitting large amounts of carbon dioxide to the atmosphere commits the planet to warm temperatures for a very long time. Once the temperatures rise, reducing emissions will not bring the temperature back down quickly. We can therefore think of climate change as being irreversible over any time period that we conceivably care about. This also means that actions we take today to curb emissions over this century will essentially determine the climate in the Year 3000 and well beyond.

The irreversibility of carbon dioxide emissions can be usefully contrasted with the second most important greenhouse gas, methane. Methane has an atmospheric lifetime of 10 years, meaning that a few decades after emitting a slug of methane,

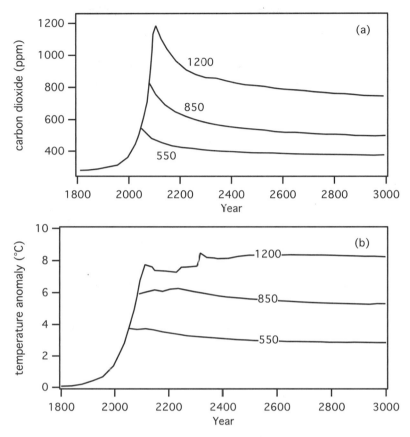

Fig. 8.6. (a) Amount of carbon dioxide in the atmosphere as a function of time, for the next 1,000 years. Carbon dioxide emissions rise at 2% per year until it hits a peak abundance (550, 850, and 1,200 ppm); then emissions are decreased instantly to zero. (b) The temperature time series corresponding to each carbon dioxide time series (adapted from Solomon et al., 2009, Fig. 1).

just about all of it is gone from the atmosphere. Thus, if humans ever stop emitting methane, we would be back down to its pre-industrial value in a few decades.

8.7 Is the climate predictable?

One criticism I frequently hear about climate predictions goes something like this: "We cannot predict the weather next week, so why does anyone believe predictions of the climate in a hundred years?" This may sound reasonable, because it is based on the correct observation that weather predictions are only accurate a few days into the future. However, the argument is built on a fatal flaw – it makes the mistake of equating weather predictions with climate predictions. In fact, it *is* possible to predict the climate in 100 years even if weather is only predictable for a few days.

The root cause of this conundrum is that predicting the weather and predicting the climate are fundamentally different problems. A weather forecast is a prediction of the exact state of the atmosphere at an exact time: "At 8 a.m. tomorrow, the temperature in Washington, D.C. will be 3 °C and it will be raining." If you get the time of an event wrong, for example, you predict rain for 8 a.m. but it does not rain until 6 p.m., then you've blown the forecast. If you predict rain for the Washington, D.C. area but the rain falls 50 km to the west in Northern Virginia, then you've blown the forecast – or if snow falls instead of rain, then you've blown the forecast.

A climate prediction, in contrast, does not require predicting the exact state of the atmosphere at any particular time; instead, it requires predicting the *statistics* of the weather over time periods of years. Thus, a climate prediction for the month of March for the years between 2080 and 2090 for a particular location might be as follows: average monthly temperature of 12 °C, with an average high of 16 °C and an average low of 5 °C; monthly average precipitation of 6.0 cm; and so on.

Being unable to make a prediction of the exact state of a complex system (e.g., the weather) does not preclude the ability to predict the statistics of the system (e.g., the climate). A coin provides a good illustration of this effect. It is virtually impossible to predict the outcome of a single flip of a coin. However, the statistics of coin flips are trivially calculable: If you flip a coin 100 times, I can tell you that you'll get approximately 50 heads and 50 tails. In other words, the inability to accurately predict any single coin flip does not preclude the ability to predict the long-term statistics of the coin.

To make this point more concretely, answer the following question: "Is it going to be hotter in Texas next January or next August?" If you know Texas weather, you would be able to predict with 100% certainty that August is the hotter month, and you can make this prediction months, years, or decades in advance. Think about that for a minute: You just made a climate prediction that is valid years in advance – far beyond the ability to predict weather.

More technically, weather forecasts belong to a class of problems known as initial value problems. This means that, to make a good prediction of the future state of the system, you must know the state of the system now. If you have a bunch of marbles rolling down a slope, and you want to predict where they'll be in one second, then you need to know where they are now to make that prediction. Similarly, to make a good weather forecast for tomorrow, you have to accurately know the state of the atmosphere today. The state of today's atmosphere is then put into a forecast model, which turns out a prediction of tomorrow's atmosphere. However, small errors in our knowledge of today's atmosphere grow exponentially, so that a forecast more than a few days in the future is dominated by the errors in our knowledge of today's atmosphere. That's why weather forecasts are not accurate for more than a few days.

Climate forecasts are boundary value problems, which is a different class of prediction that does not require knowledge of today's atmospheric state. Instead it requires a knowledge of the radiative forcing of the climate. This is why, for example, we can predict with 100% certainty that August in Texas will be on average hotter than January in Texas. We know this because we know that more sunlight falls on Texas and the rest of the northern hemisphere at that time, leading to higher temperatures.

Increases in greenhouse gases also increase the heating of the surface, although by infrared radiation instead of visible. Thus, we can have confidence that, if we add enough greenhouse gas to the atmosphere, the increase in surface heating will warm the planet – just as we can predict that summer will be hotter than winter.

One should not take this to mean that predicting the climate is an easier problem than predicting the weather, only that they are different problems. Some aspects of the climate problem are, in fact, harder than the weather problem. For example, because weather forecasts cover only a few days, weather models can assume that the world's oceans and ice fields do not change. Climate models, however, cannot make this assumption, because both the world's oceans and ice fields can significantly change over a century. Climate models must therefore predict changes in these and other factors in order to accurately predict the evolution of the climate system over a century.

8.8 Chapter summary

- Prediction of future climate requires predictions of future emissions of greenhouse gases from human activities. Such predictions are known as emissions scenarios.
- The factors that control emissions are population (P), affluence (A), and greenhouse-gas intensity (T). This is expressed in what is known as the IPAT relation: $I = P \times A \times T$, where I is carbon dioxide emissions.
- Greenhouse-gas intensity is the product of energy intensity and carbon intensity. Energy intensity reflects the efficiency with which the society uses energy as well as the mix of economic activities in the society. The carbon intensity reflects the technologies the society uses to generate energy.
- Because predictions of the future are so uncertain, the IPCC has put forth a set of plausible, alternative scenarios of how the world might evolve. Taken as a group, these scenarios span the likely range of future emissions trajectories.
- Putting these emissions scenarios into a climate model yields predictions of warming over the 21st century of 1.8–3.6 °C. This is several times the warming of 0.7 °C that the Earth experienced over the course of the 20th century.
- Climate change does not stop in Year 2100. Carbon dioxide stays in the atmosphere for centuries after it is emitted, so elevated temperatures will last for a very long time after emissions stop. Large emissions during this century will cause the Earth's temperatures to remain elevated for this millennium and beyond.

Additional reading

IPCC, Special Report on Emissions Scenarios, 2000. This describes the emissions scenarios discussed in this chapter in great detail. Report available on-line at: http://www.ipcc.ch/pdf/special-reports/spm/sres-en.pdf.

For a more recent set of emissions scenarios, see the U.S. Climate Change Science Program, *Scenarios of Greenhouse Gas Emissions and Atmospheric Concentrations (Part A) and Review of Integrated Scenario Development and Application*

(Part B). A Report by the U.S. Climate Change Science Program and the Subcommittee on Global Change Research [Clarke, L., J. Edmonds, J. Jacoby, H. Pitcher, J. Reilly, R. Richels, E. Parson, V. Burkett, K. Fisher-Vanden, D. Keith, L. Mearns, C. Rosenzweig, M. Webster (Authors)]. (Washington, DC: Department of Energy, Office of Biological & Environmental Research, 2007). Report available on-line at: http://www.climatescience.gov/Library/sap/sap2-1/finalreport/default.htm.

Efforts to improve emissions scenarios continue. For a description of the scenarios that will be used in the next IPCC report, see R. H. Moss et al., "The Next Generation of Scenarios for Climate Change Research and Assessment," *Nature* 463 (2010): 747–756 (doi:10.1038/nature08823).

D. Archer, *The Long Thaw: How Humans Are Changing the Next 100,000 Years of Earth's Climate* (Princeton, NJ: Princeton University Press, 2010). Among the many things covered in this book is the very long-term evolution of climate change.

Terms

Carbon intensity
Emissions scenario
Energy intensity
Greenhouse-gas intensity
Gross domestic product (GDP)
IPAT relation

Problems

1. a) Someone asks you about how much the climate will warm over the next 100 years if we do nothing to address climate change. How do you answer?
   b) If the amount of carbon dioxide and other greenhouse gases stopped increasing today and were held constant into the future, how would the climate evolve over the next century?

2. a) Define each term in the IPAT identity.
   b) What are the units of each term? Show how the units cancel so that the *I* term has units of emissions of greenhouse gases.

3. a) The *T* term can be broken into two terms. What are these two terms, and what are their units?
   b) If we switch from fossil fuels to solar energy, which of the terms change, and does this term increase or decrease?
   c) If we convert from traditional incandescent lighting to LED lights, which of the terms change, and does this term increase or decrease?
   d) If we switch from natural gas to coal, which of the terms change, and does this term increase or decrease?

4. Explain the step-by-step process by which future predictions of climate are generated.

5. Consider this argument: "We cannot predict the weather in a week, so there's no way we can believe a climate forecast in 100 years." Is this argument right or wrong? Explain your answer.

6. If we emit significant amounts of carbon dioxide this century, how long will the planet remain warm?

7. a) What are the names of the four main emissions scenarios discussed in this chapter?

   b) In just a few sentences, explain the main differences between them.

8. Explain how your level of wealth impacts how much emission of carbon dioxide you are responsible for.

9. In 2002, the Bush Administration set a goal of reducing greenhouse gas intensity by 18% by Year 2012. Would achieving this goal result in a decrease in emissions?

9

Impacts

Before the summer of 2010:

> Russia is a northern country and if temperatures get warmer, it's not that bad. We could spend less on warm coats.
>
> – Vladimir Putin, President of Russia[1]

After the summer of 2010:

> Practically everything is burning. The weather is anomalously hot . . . What's happening with the planet's climate right now needs to be a wake-up call to all of us, meaning all heads of state, all heads of social organizations, in order to take a more energetic approach to countering the global changes to the climate.
>
> – Dmitri Medvedev, President of Russia[2]

Warmer temperatures might not sound too bad – you might associate them with fun things such as vacations at the beach or summer cookouts. The reality is quite different. We rely in very important ways on the stability of the climate – for things such as food and fresh water. Most people just don't notice this reliance because it has been obscured by two centuries of scientific, technological, and economic advancements.

Every once in a while, however, an event comes along that reminds us of the impact of climate on our lives. The heat wave of the summer of 2010 was one of those events for the Russians. The Russians learned the hard way that warmer temperatures do not mean tank tops and grilled hot dogs, but instead mean wildfires, loss of agricultural crops, and human suffering.

Nonetheless, you might have wondered as you read the first eight chapters, "So the climate is changing. Why should I care?" In this chapter, I address this question. By the end of the chapter, I hope that you recognize that climate change is a significant risk that we ignore at our peril.

9.1 Why should you care about climate change?

In Chapter 8, we saw that if the world does nothing to address climate change, we can expect global average temperatures to increase by a few degrees Celsius during

[1] Quoted in "Nyet to Kyoto, Blow for Campaign as Putin Jokes about Global Warming," *The Mirror*, September 30, 2003, p. 4.

[2] Quoted in "Will Russia's Heat Wave End Its Global-Warming Doubts?," *Time Magazine*, August 2, 2010.

the 21st century. This may not seem like much warming to you. After all, in many places the summer days are 50 °C warmer than the winter days and daytime can be 25 °C warmer than the subsequent night. And one day can be several tens of degrees Celsius warmer or cooler than the next. If you consider the size of these temperature variations, a change in the global average of a few degrees may sound insignificant.

In this case, however, your intuition is wrong. Although the temperature in any single place can vary considerably by season, by day, and even within a day, the variations tend to cancel when averaged over the entire globe. When you are experiencing the warmth of daytime, someone on the other side of the globe is experiencing the coolness of night. When it is summer where you live, it is winter in the other hemisphere. Heat waves in one location are generally canceled by a cold spell somewhere else (e.g., Figure 2.1). As a result, the temperature variations you experience are nearly completely canceled by opposite variations somewhere else on the Earth.

Because of this cancellation, the global average temperature of the Earth is very stable, as we can see in Figure 2.2. Moreover, seemingly small changes in global average temperature are associated with significant shifts in the Earth's climate. For example, the global annual average temperature during the last ice age was 5–8 °C colder than that of our present climate. At that time, the Earth was basically a different planet: Glaciers covered much of North America and Europe, and because so much water was tied up in glaciers, the sea level was approximately 120 m lower than it is today.

And during the summer of 2003, a heat wave struck Europe and led to the deaths of several tens of thousands of people. During that heat wave, the average temperature in Europe was only 3.5 °C above the average of the past century. In addition, temperatures today are perhaps 1 °C warmer than they were a few hundred years ago, a period whose climate is different enough that it's been dubbed the Little Ice Age. Thus, we should take projections of a few degrees of warming very seriously.

Furthermore, it's not just the size of the warming but the rate of warming that is of concern. It took more than 10,000 years for the planet to warm 5–8 °C and emerge from the last ice age – a rate of less than one tenth of a degree per century. Moreover, warming since the Little Ice Age is approximately 1 °C in a few hundred years. The prediction of 21st-century warming of a few degrees Celsius over just one century is incredibly fast in comparison.

The rate matters because the faster the warming occurs, the less time people and the environment have to adapt to the changes. If the sea level rises by 1 m in 1,000 years, it seems likely that we could adapt gracefully to that change. But a 1 m increase in sea level in a century would be much harder to adjust to. And a 1 m increase in a decade would be nearly impossible to adapt to – it would displace millions of people and destroy trillions of dollars of infrastructure.

Another argument often made is that a warming of a few degrees should not cause concern because the Earth has gone through such warmings and coolings many times during its 4-billion-year history. This is undoubtedly true, as was discussed in Section 2.2. However, modern human society, with a population of several billion people, metropolitan areas with tens of millions of people, and reliance on industrial farming and large-scale built infrastructure is only a century or two old. Over this time, the Earth's climate has been stable, varying by less than 1 °C. Modern human society has never had to face several degrees of warming in a century. Thus, the argument that we've experienced this type of warming before is simply wrong.

Finally, you might be asking, "How do I know that a warmer climate won't be better?" The reason it won't be is because both human society and the natural world are adapted to our present climate. Humans in particular have built their world based on the climate that exists right now. If the climate changes, then we will necessarily be less well adapted to our environment. As an analogy, imagine that you go to the tailor and get a suit fitted exactly to the shape of your body. At that point, no change in your body shape will improve the way the suit fits – for example, either gaining or losing weight will cause the suit to fit less well. In a similar fashion, virtually any change in the climate – warming or cooling – will result in overall negative outcomes for human society.

As an example, consider that some communities in Alaska are built on permafrost. This causes no problem as long as the permafrost remains frozen, but if it warms and melts then it can destroy whatever is built on it – buildings, roads, bridges, and so on. Thus, when you build on permafrost, you're assuming that the climate will remain stable enough to keep the permafrost frozen. Unfortunately, Alaska is experiencing rapid warming, and melting permafrost is causing damage to the State's infrastructure.

Agriculture provides another example of adaptation to our present climate. We farm where the climate provides suitable growing-season temperature and precipitation. Around these farmers we build essential infrastructure to support agriculture: grain silos, processing plants, tractor dealers, seed suppliers, and the like. If the climate shifts and the temperature and precipitation are no longer conducive to farming, then all of these investments in the agriculture infrastructure will no longer useful. We will then have to abandon the infrastructure that exists and rebuild it in whatever region becomes conducive to farming.

As a final example, imagine that you build a marina on the edge of a lake. As soon as you pour the foundation, you are optimized for that particular lake level. If the lake level goes up, your marina floods. If the level of the lake drops, then you also face problems. If the drop is not too much, then you can make small changes such as extending the piers and boat ramps. If the lake drops enough, though, you must give up the original site and rebuild the marina closer to the new edge of the water. In the Western United States, where water levels are dropping in response to warmer temperatures, increased demand, and drought, this process of following the receding edge of the lake even has a name – "chasing water."

The upshot of this discussion is that, when it comes to climate, *change is bad*.

In this chapter, I will break the climate impacts into two components. The first set includes the physical impacts on the climate system: how temperature, precipitation, sea level, extreme events, and other such phenomena will change. The second is the impact of such changes on humans and those aspects of the environment that we rely on and care about. I will address these two issues in turn.

I should also point out upfront that there is significant uncertainty in specific estimates of future impacts of climate change. Such estimates require accurate predictions of climate at regional scales, as well as an understanding of how these changes will

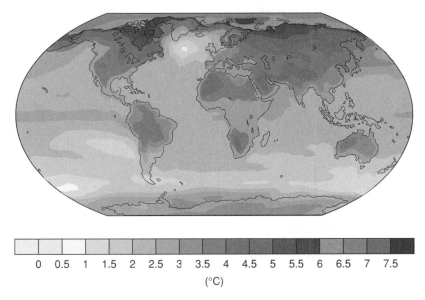

0   0.5   1   1.5   2   2.5   3   3.5   4   4.5   5   5.5   6   6.5   7   7.5

(°C)

Fig. 9.1   Increase in average surface temperature between the end of the 20th century (1980–1999) and the end of the 21st century (2090–2099). This is calculated from an ensemble of climate models driven by the A1B scenario, a mid-level emissions scenario (the source is IPCC, 2007a, Fig. SPM.6). (See Color Flate 9.1.)

affect humans and how humans will respond and adapt to the new climate. Nonetheless, although it is impossible to say exactly how climate change will impact any particular region or person, there are some general conclusions that we do have high confidence in. These things will be the main focus of this chapter. In Chapter 14, I will discuss the implications of the pervasive uncertainty on the policy debate.

9.2 Temperature

Although the global average temperature of the planet is currently increasing and is almost certain to continue warming from each decade to the next, the increase is not uniform across the globe. Figure 9.1 shows a prediction of the distribution of warming over the 21st century, and there are some obvious features in the warming distribution. First, continents warm more than oceans because of the larger heat capacity of the oceans. As a result, warming in northern North America and Eurasia is projected to be more than 40% greater than the global average warming. Second, high latitudes will warm more than the tropics. This is primarily due to the ice–albedo feedback: The warming causes loss of ice, and the loss of ice exposes dark ocean, which absorbs more sunlight and leads to further warming. The models also predict more warming in the Arctic than in the Antarctic. These changes are all continuations of trends that have been observed over the past century (e.g., Figure 2.2b).

Not apparent from this plot is that adding greenhouse gases to the atmosphere tends to reduce temperature contrasts. Thus, we expect more warming in winter than in summer (reducing summer–winter temperature contrasts), and more warming at night than during the day (reducing day–night temperature contrasts). To understand this result, you need to recognize that temperature variations in our climate are generally caused by variations in the distribution of sunlight (e.g., the high latitudes, nighttime, and winter are all colder because they receive less sunlight than the tropics, daytime, and summer). The atmosphere also heats the surface, but because greenhouse gases tend to be well mixed in our atmosphere (because of their long residence times), the heating from greenhouse gases occurs evenly over the entire surface of the planet, and it is the same at night as during the day and the same in winter as in summer.

As the abundance of greenhouse gases increases, heating of the surface from greenhouse gases becomes stronger while heating from sunlight remains about the same. Thus, variations in solar heating with latitude, time of day, and season have less of an impact on total heating of the surface. As a result, these temperature contrasts will decrease. In the limit of a planet such as Venus, with a massive greenhouse-gas-rich atmosphere, the heating of the surface by the atmosphere is roughly 16,000 W/m^2, which is a hundred times more than solar heating. As a result, the variations in the solar input are comparatively miniscule and the temperature everywhere on Venus is approximately the same: 735 K.

Figure 9.2 shows a close-up of the United States. Most of the warming is in the middle of the continent, far from the moderating influence of the oceans, and at high latitudes. In 2050, warming in the high-emissions scenario is 2–3 °C while warming in the low-emissions scenario is 1–2 °C. The small difference in 2050 between high- and low-emissions scenarios reflects the large inertia in our climate system – both physical, because of the large heat capacity of the oceans, as well as inertia in our ability to shift to nonemitting technology. As a result, the warming over the next few decades is essentially already determined by past emissions and our present mix of energy technology.

In 2090, however, the warming for the high-emissions scenario is 4–5 °C, which is roughly twice the warming in the low-emissions scenario of 2–3 °C. Thus, we do still have significant control of warming in the second half of the 21st century.

In addition to increases in average temperatures, it is also very likely that extreme temperature events will also change. Figure 9.3 shows that, for a high-emissions scenario climate, a day so hot that it currently occurs once every 20 years will occur at least every other year by the end of the century over most of the United States. Of course, this also means that cold days and cold waves will become less frequent.

Overall, we can have high confidence in the general shape of these predictions. There is, of course, uncertainty in the exact magnitude of the warming, both from uncertainty in the physics and uncertainty in the emissions pathway, but it is highly likely that the temperature is going up every decade and that the general distribution of the warming will be in accord with these model predictions.

9.3 Precipitation

As greenhouse gases increase, E_{in} for the surface increases because of increased infrared radiation from the atmosphere falling on the surface. This leads to an increase

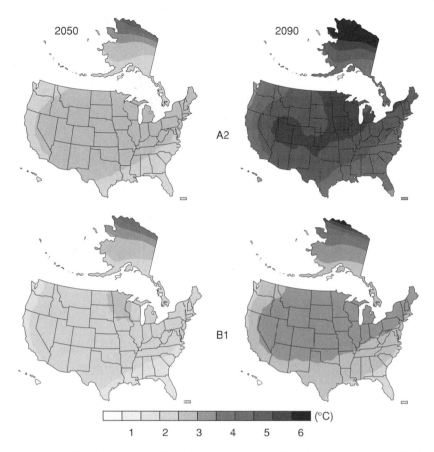

Fig. 9.2 Annual warming of the United States in 2050 (left-hand column) and 2090 (right-hand column) relative to the 1961–1979 average for the low-emissions scenario B1 (bottom row) and the high-emissions scenario A2 (top row). These are calculated from an ensemble of climate models (the source is Karl et al., 2009, p. 29). (See Color Plate 9.2.)

in evaporation from the oceans, and because precipitation must balance evaporation, this means that precipitation also increases. More quantitatively, this translates into total global precipitation that increases by a few percent for every degree Celsius of global average warming.

Although total rainfall is expected to increase, the increase will not be distributed evenly. We expect precipitation to increase in the high latitudes and decrease in most subtropical land regions as well as some parts of the tropics. Predictions of changes in annual average precipitation from a set of climate models are shown in Figure 9.4.

In addition, continuing a trend of the 20th century, it is likely that more of the rainfall will come in the heaviest downpours. During a heavy downpour, the soil saturates before the end of the rain event, and the remaining rain therefore runs off. Heavy runoff can lead to a number of negative consequences, such as increased erosion and a higher risk of flooding.

Because more water runs off during a heavy rain event, an increase in the fraction of rain falling in heavy rain events means that there will be less water available

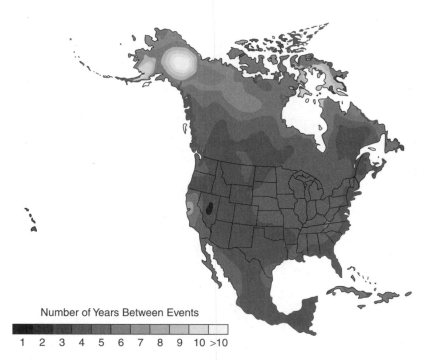

Fig. 9.3 Model predictions, calculated by using the high-emissions (A2) scenario, of the time interval between extremely hot days in 2080–2099 that today occur every 20 years (the source is Karl et al., 2009, p. 33). (See Color Plate 9.3.)

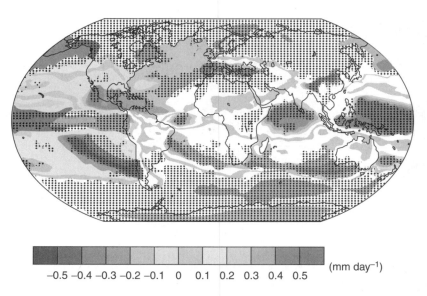

Fig. 9.4 Change in annual mean precipitation over the 21st century as predicted by climate models driven by a mid-range emissions scenario (A1B). Stippled regions are where at least 80% of models agree on the sign of the mean change (data obtained from Meehl et al., 2007, Fig. 10.12). (See Color Plate 9.4.)

to use by the ecosystem. It also tends to increase the time between rain events. Combined with warmer summers, which will increase the rate at which water is lost from soils by evaporation, this will increase the frequency and intensity of drought. Thus, we get the surprising result that both wet and dry extremes will grow more likely in the future: wet extremes, with associated risks of flooding, increased erosion, and landslide; and dry extremes, with associated risks of water shortages and drought.

In addition to changes in the amount of precipitation, there will also be shifts in the form. Less wintertime precipitation will fall as snow and more will fall as rain. As I will discuss shortly, snow has the property that it does not run off until it melts, so changing the form of precipitation will change the timing of runoff, which has important implications for water availability.

Unfortunately, these predictions of future precipitation patterns are quite uncertain. Precipitating clouds can be small (sometimes only a few kilometers across), and climate model grids are too coarse to simulate such small entities. As a result, precipitation must be parameterized in climate models. Model predictions based on parameterized aspects of the atmosphere should be considered to be more uncertain than those that the climate model explicitly resolves. This is why weather forecasts, for example, incorrectly predict precipitation far more frequently than temperature. Thus, although we have high confidence in the general predictions at a qualitative level (e.g., shifts in precipitation location, more rain falling in heavy events), we have less confidence in the exact details of the changes, such as exactly which regions will see changes, and how big those changes will be.

9.4 Sea-level rise and ocean acidification

Sea-level rise is one of the most certain impacts of climate change. As we learned in Chapter 2, the sea level rises in response to warming temperatures for two reasons. First, as grounded ice melts, the melt water runs into the ocean, increasing the total amount of water in the ocean and, therefore, sea level. Second, like most things, water expands when it warms. Measurements (Figure 2.8) confirm that sea levels have indeed been rising as temperatures have gone up, and we can be certain that the seas will continue to rise into the next century.

The 2007 IPCC report predicted increases in sea level of several tens of centimeters over the next century. However, this estimate excluded rapid changes in the world's big ice sheets on Greenland and Antarctica, so it should be considered a lower limit. More recent work suggests that we can expect at least 1 m of sea-level rise this century unless emissions are significantly reduced.

Ocean acidification is another certain consequence of continued emissions of carbon dioxide. As we explored in Chapter 5, a significant fraction of carbon dioxide emitted to the atmosphere by humans ends up in the oceans. In the liquid environment of the ocean, carbon dioxide reacts with water and is converted into carbonic acid, the same weak acid found in soda. The net result is that, as the oceans absorb more and more carbon dioxide, the oceans will become more acidic.[3] In fact, since the industrial revolution, the absorption of carbon has lowered the ocean's pH by approximately 0.1. At present, the ocean is now more acidic than it has been for more than 20 million years.

9.5 Impacts of these changes

The physical changes in the climate system described above are only the first step in determining climate impacts. Determining impacts also requires an understanding of our sensitivity to climate change – in other words, how climate change will affect us, either adversely or beneficially, as well as an understanding of how human responses can ameliorate their impact.

Humans are directly sensitive to rising temperatures, particularly extreme heat events, as evidenced by the tens of thousands of people who perished during the European heat wave in the summer of 2003. In fact, extreme heat is the greatest weather-related cause of death in the United States. Thus, increases in temperatures, particularly extreme events, will produce negative health consequences, with the extent of these negative impacts determined by exactly how much warming occurs, how it is distributed, and how people adapt to the warmer temperatures. It should also be noted that there are also some benefits to warmer temperatures, such as reductions in the negative impacts of cold temperatures and extreme cold events. However, the benefits of less frequent cold temperatures are not equal to the costs of more frequent high temperatures.

Warmer temperatures also affect other things we care about, such as agriculture. During the European heat wave, for example, there was a one-third decrease in the yields of grains and fruits. This occurs because (among other reasons) photosynthesis in some of our most important crops is most efficient at temperatures between 20 °C and 25 °C, with yields declining precipitously at temperatures much above 30 °C. Even under the most favorable emissions scenario (B1), yields of our most important crops are expected to decline by roughly one third or 33% over the next century, whereas under the most pessimistic emissions scenario (A1FI), yields are expected to decline by 70%.

In the public debate, an argument frequently heard is that increasing carbon dioxide will be good for plants – that is, "carbon dioxide is plant food." Like many arguments in the public climate change debate, this one is partially true. Atmospheric carbon dioxide is indeed a key ingredient in plant growth and, everything else being equal, more carbon dioxide in the atmosphere would be expected to increase the rate of plant growth. However, not everything else is equal. As the amount of carbon dioxide in the atmosphere increases, other things change – such as temperature and

---

[3] The present pH of the ocean is approximately 8, meaning that it is a base. Acidification here means that the pH is decreasing, not that the ocean will actually become acidic. For the ocean to actually become acidic, its pH would have to drop below 7, which is very unlikely.

precipitation. Although some regions may see benefits to agriculture, the research that's been done on this suggests that the beneficial effects of increased carbon dioxide are overwhelmed by the negative effects of these other changes, particularly if the warming is severe.

Many systems managed for human use, such as agriculture, commercial forests, rangelands, and fisheries, will be affected by climate change. Because human management dominates these systems, on the one hand, disruption of these systems by climate change may have severe human impacts because we depend on them so much. On the other hand, the ability to adapt management practices to changing conditions offers the possibility of mitigating at least some of the harmful impacts.

Warmer temperatures also affect ecosystems that are not directly managed by humans. For example, new research shows that warming temperatures are presently driving lizards to extinction. During spring, when energy demands are highest because lizards are reproducing, the warming temperatures reduce the amount of time that lizards can forage for food (if the temperatures get too high, cold-blooded lizards have to rest). If temperatures continue to increase (which we expect), then at some point the time available to look for food diminishes to the point where lizards simply cannot find enough food – and extinction ensues. This is already happening, and extrapolating into the future, we see that global warming may lead to extinction of 40% of all global lizard populations by 2080. Note that it is not the global average temperature that matters to lizards, or even the local average temperature – but the local daily temperatures during one particular time of the year. This shows that it is the details of climate change that ultimately matter, not the broad-brush changes in global average quantities.

Now you may not care much about lizards, but you should care about their extinction for two reasons. First, the environment is a tightly coupled system. There are many examples in history in which humans have intentionally removed a species from the environment because they thought it was harming them (e.g., getting rid of birds because the birds were eating crops) only to find out that that change led to more problems than it solved (e.g., the birds are also eating insects, and with the birds gone the insects proliferated and destroyed much more of the crops than the birds were eating). Today's modern world obscures many of these relationships, but they still exist. Removing lizards from an ecosystem may have important effects on the rest of the environment that we do care about, just like pulling a single thread on a sweater can unravel the entire thing.

Second, this is not just about lizards. As warming temperatures drive lizards to extinction, the same warming temperatures will be having similar deleterious effects on many other species. In fact, a significant fraction of plant and animal species may be at increased risk of extinction if global average temperatures increase by a few degrees Celsius. Thus, at the expense of a mangled metaphor, lizards may be the canary in the coalmine.

In addition to impacts on individual plant and animal species, we also need to consider the impact on entire ecosystems, such as alpine meadows or temperate forests. As the climate changes, ecosystems will not remain intact. Rather, each component species of the ecosystem will be affected in its own way. Some species may adapt readily, whereas others may be unable to adapt fast enough to survive. Species will also be subject to human interventions and constraints such as land-use change, barriers, and intentional or inadvertent transport.

The aggregate result will be that ecosystems will evolve, with new relationships among incumbents and new arrivals developing in each location. In some cases, the new assemblies may be similar enough to present ecosystems that we can think of them as basically unchanged. In other cases, however, the new systems may be unlike any present ecosystems, with new species and relationships between them and other ecological surprises. Some ecosystem types are likely to be lost entirely, such as alpine systems, coastal mangrove systems, and coral reefs. The consequences to humans from the loss of the services provided by ecosystems may be substantial. Mangrove systems, for example, provide important protection for coastal areas from erosion, storm surge (especially during hurricanes), and tsunamis. Their loss makes us more vulnerable to these occurrences.

Loss of sea ice is another important impact of rising temperatures. Measurements (Figure 2.5) show that the area covered by sea ice has been decreasing over the past few decades in the Arctic as that region has warmed. So has ice thickness, resulting in a large net decrease in Arctic ice volume. Antarctic ice has not been decreasing, which is consistent with less warming found there (Figure 2.2b).

In the future, we can expect sea ice to continue to decrease in the Arctic, and to begin to decrease in the Antarctic as temperatures there begin to increase. Most predictions are that the Arctic ocean will be entirely ice free during summertime at some point in the 21st century, although exactly when is still being debated.

Less sea ice is expected to adversely affect Arctic ecosystems and wildlife, such as polar bears and seals. In addition, sea ice is a crucial part of the culture of those individuals who live in the Arctic, and the loss of it will irreversibly change the way those people live. However, there are advantages to the loss of sea ice, such as faster maritime transport of goods between Europe and Asia, as well as increased access to Arctic resources.

Changes in precipitation, combined with increasing temperature, are expected to lead to changes in the availability of fresh water. In some places, overall water availability is expected to increase. For example, river runoff is projected to increase by 10% to 40% by mid-century at higher latitudes and in some wet tropical areas, including populous areas in East and Southeast Asia.

However, the beneficial impacts of increased annual runoff in these areas may be tempered by changes in the timing of the runoff. For example, summertime runoff from melting snowpack provides an important source of fresh water to the U.S. Pacific Northwest at a time when there is little rainfall. Warming temperatures, however, will lead to less wintertime precipitation falling as snow and more as rain, and the snow that does fall will melt earlier. Both of these effects will tend to shift runoff from the summertime, when the water is most needed, toward winter and spring, when it is less needed. This will increase stresses on summertime water supplies. In addition, the shift of precipitation to fewer, heavier events will also lead to further changes in water availability.

In the mid-latitudes and dry tropics (e.g. the Mediterranean Basin, Western United States, southern Africa, and northeastern Brazil), decreases in rainfall and increases in temperature will lead to a significant decrease in water resources. Drought-affected areas are projected to increase in extent, and bring with them adverse impacts on many things that we care about, such as agriculture, water supply, energy production, and health.

Fresh water may be the single most critical resource that humans need. The most important use of fresh water is in agriculture – it takes 1,000 tons or so of water to produce 1 ton of grain, and 15 times as much to produce 1 ton of meat. Reductions in the availability of fresh water will therefore cause problems for food production. In addition, fresh water plays important roles in many industrial applications. For example, most plants that generate electricity require huge amounts of water for cooling, which is why these plants are usually situated on rivers or lakes. During droughts, low water levels sometimes force these plants to shut down. Overall, the negative impacts of climate change on systems and activities reliant on fresh water are generally expected to be significant.

Rising seas will also bring impacts that are strongly negative because a significant fraction of the world's population lives within a few feet of sea level. Moreover, some of the world's most productive farmland is located in river deltas and other regions that are particularly sensitive to sea-level rise. Even small amounts of sea-level rise will therefore have significant negative implications. In Florida, for example, a sea-level rise of 68 cm (which is likely to occur in less than 100 years) would inundate 9% of Florida's current land area at high tide.[4] This includes virtually all of the Florida Keys as well as 70% of Miami-Dade County. Almost one tenth of Florida's current population, or 1.5 million people, live in this vulnerable zone, and it includes residential real estate now valued at over $130 billion. It also includes important infrastructure, such as two nuclear reactors, three prisons, 68 hospitals, and so on. And this is just Florida. Multiply these impacts by a few hundred times to account for all of the places on the planet where people live near sea level, and you can get a feel for how big a problem this is going to be.

And those regions not actually submerged will also be affected. Increased sea level will increase the frequency of flooding from extreme sea-level events, so that a flood event that occurred, say, every 100 years, may occur after sea-level rise every few years. As the flooding of New Orleans after Hurricane Katrina showed, these events cause significant loss of life as well as economic destruction.

It's worth noting that our sensitivity to sea level is a particularly good example of how we are adapted to our present climate. When cities such as New Orleans or Miami were founded, the original inhabitants did not take into account the possibility of sea-level rise. Rather, they assumed that the sea level would remain pretty much as they found it. Because of this, any change in sea level, either up or down, would have negative effects on these cities.

The acidification of the ocean is another important impact. Organisms often build shells or skeletons out of calcium carbonate, and their ability to do this will be strongly affected by increases in acidity. At first, these species will find it more difficult to extract carbonate from the water for use in their shells or skeletons. Eventually, the acidity increases to the point where it is fatal for the species. It is

---

[4] See Stanton and Ackerman (2007).

important to realize that ocean acidification is not just a theory – it has happened before. During the PETM (discussed in Section 7.5), a massive amount of carbon was emitted into the atmosphere, which subsequently dissolved into the ocean. That event was accompanied by an acidification of the ocean that dissolved much of the carbonate sediment there (Figure 8.6).

Some species will adapt better to increasing acidity than others. As a result, the mix of species in ocean ecosystems will shift, resulting in new and novel arrangements of species. Such changes could have important consequences for species that rely on the ocean for food – including us.

We can also expect human health to be negatively impacted by climate change. Some of these health impacts follow closely from changes already discussed, such as negative health consequences of warmer temperatures or malnutrition associated with reductions in food availability. In addition to those impacts, we expect warmer, more humid days to enhance the photochemical reactions that cause air pollution, leading to more smoggy days as the climate warms, along with the associated health impacts of air pollution.

Warming temperatures also increase disease risk as a result of expansions in ranges of animals that transmit the diseases (e.g., mosquitoes), shortening of the diseases' incubation periods, lack of very cold temperatures that can kill the transmitters, and disruption and relocation of large human populations. Moreover, increases in water temperature, precipitation frequency, and other factors could increase the incidence of water contamination with harmful pathogens, resulting in increased human exposure.

Finally, national defense analysts at the Pentagon have been thinking about the impact of climate change on national security. They worry, for example, that climate impacts will destabilize societies through the disruptions of economic activity, food production, and water availability, as well as other mechanisms. At the minimum, these disruptions will lead the most affected people to pack up and try to move to a better place to live. For example, a mega-drought hitting Central America might lead to a surge of environmental refugees heading toward the United States, which would cause various security, social, and economic disruptions. At worst, these disruptions could lead to social destabilization, failed states, and war.

But humans are smart, and as the climate changes, people have the opportunity to make changes that will minimize the impacts of climate change. For example, many buildings in Boston are currently not air conditioned, because there are only a few days a year that are hot enough to require cooling. As the climate warms, air conditioners can be installed in buildings as needed to accommodate the heat. Similarly, increases in sea level can be dealt with by building seawalls or by relocating people.

Dealing with the impacts of climate change in this way, however, takes significant resources. This means that the ability to adapt is not evenly spread across the globe. Rich, well-governed places such as the United States or Europe have resources that can be applied to adapting to climate change. For small climate change (the bottom end of the range in Figure 8.5), these countries will likely find most effects of climate change to be manageable without too much social disruption. If climate change falls toward the upper end of the predicted range in Figure 8.5, more than 2–3 °C or so above pre-industrial temperatures, then climate change is expected to be a serious, perhaps insurmountable challenge for even these rich countries.

Approximately 2 billion people, however, are so crushingly poor that they have no additional resources available to address climate change. For them, installing air conditioners or building coastal defenses in response to rising seas, developing new freshwater infrastructure in response to water shortages, improving public health infrastructure in response to a new disease outbreak, and the like are simply not affordable options. For these poorest of the poor, even small climate change will be a major challenge. Furthermore, if climate change falls toward the upper end of the predicted range in Figure 8.5, then it would be a certain disaster for the poorest. I will return to this issue when I talk about adaptation in Chapter 11.

A particularly important factor in determining the severity of the impacts will be the rate of climate change. Ecosystems have adapted to large climate change in the past, such as the warming from the previous ice age. However, the warming predicted for the next century will be as much as 50 times faster. As the rate of warming goes up, the ability of the environment to gracefully adapt to the changes declines. What is uncertain is how much less gracefully, and with what consequences.

9.6 Abrupt climate changes

The changes I just described – changes in temperature, precipitation, and sea level – are all certain. These changes *will* be happening; the only question is the magnitude and distribution of the changes, and how well humans and the natural environment cope. But there is also the possibility of a more drastic and consequential impact occurring: an abrupt climate change.

In this context, an abrupt climate change is a sudden and significant shift in some aspect of the climate. As an analogy, imagine that you're sitting in a canoe and you start to lean over. At first, the canoe tilts with you – until, that is, you pass a critical threshold and the canoe suddenly flips over, throwing you and everything else in the canoe into the river. That's an abrupt change.

For the climate, the worry is that the climate will not warm smoothly as greenhouse gases are added to it, as suggested by plots such as Figure 8.5. Rather, we will add enough greenhouse gas that the climate system would undergo a large and rapid shift to an entirely new climate state – equivalent to the canoe rapidly transitioning from right side up to upside down. In the case of climate, the large climate shift might occur in just a few decades.

This possibility concerns scientists because abrupt changes have happened in the past. During the PETM approximately 55 million years ago, there was a rapid release of greenhouse gases and a subsequent warming of 5–9 °C in just a few thousand years. It's not known what caused the release of greenhouse gases, but one possibility is that it was due to a carbon cycle feedback – for example, an initial warming melts permafrost, leading to the release of carbon stored in it, which leads to more warming, and so on. If so, then such an occurrence could happen again.

In addition, roughly 12,000 years ago, as the Earth was emerging from the depths of the last glacial maximum, the temperature suddenly plunged (at least in the mid- and high latitudes of the northern hemisphere). The period of low temperatures during the millennium that followed, today known as the Younger Dryas, is thought to have been due to a massive release of water into the North Atlantic from melting glaciers. This freshwater influx disrupted the ocean currents, in particular the Gulf Stream. Because the Gulf Stream transports heat from the tropics to the high latitudes, the shutdown of the Gulf Stream caused mid- and high-latitude temperatures to plummet (this was the basic scientific premise behind the movie *The Day After Tomorrow*).

Thus, abrupt changes do happen and we must take their possibility seriously. However, beyond acknowledging the possibility, there is little the scientific community can say. Climate models do not predict the occurrence of an abrupt climate change, and most experts view the probability to be low, but not zero. If an abrupt change did occur, though, it would be an unprecedented catastrophe. Such low-risk, high-consequence events pose significant challenges to our society. The tendency is to ignore the risk until it occurs, which is why dams are built after floods, and not before. However, in this case, the likelihood is that, once the abrupt change takes place, it will be very difficult, if not impossible, to reverse. This makes the strategy of ignoring the risk a precarious proposition.

9.7 Chapter summary

- Predictions of warming of a few degrees Celsius over the next century should be taken seriously. This is a huge amount of warming; for comparison, the last ice age was only 5–8 °C colder than the temperature today.
- We are adapted to our present climate, so any significant change is likely to be detrimental.
- The warming will not be uniformly distributed. We expect the land to warm more than the oceans, and high latitudes, particularly the Arctic, to warm more than the tropics. We also expect temperature contrasts to decrease: Night will warm more than day, and winter will warm more than summer. Extreme heat events will occur more frequently and be more intense.
- Precipitation will change. Although global average precipitation is expected to increase, the more important change will be in the distribution of rainfall. We expect precipitation to increase in the high latitudes and decrease in most subtropical land regions as well as some parts of the tropics.
- Continuing a trend of the 20th century, it is likely that more of the rainfall will come in the heaviest downpours. This leads to the counterintuitive result that both wet extremes (e.g., flooding) and dry extremes (e.g., drought) will become more frequent in the future. We also expect the timing of runoff to change, which will alter the availability of fresh water.
- Sea level is going to rise and the oceans will become more acidic.
- These changes may have important negative impacts on humans. This includes impacts on agriculture and freshwater availability, as well as public health consequences. Climate change is also a destabilizing influence that has negative implications for our society.

- The impacts of these changes on human society will not be distributed evenly. The wealthy countries of the world will likely have an easier time adjusting than will the poor countries of the world, and there may be some people who benefit, particularly if the warming is small. As the amount of warming increases, negative impacts will increasingly dominate the benefits.
- Abrupt changes are low-probability, high-consequence events. An example of an abrupt change is a reorganization of the ocean's circulation. Although scientists do not expect them to occur this century, they cannot be ruled out.

Additional reading

Much has been written about the impact of climate change on humans and the rest of the environment. Here are a few notable works.

The IPCC's Working Group II focuses on impacts of climate change and our ability to adapt to them, and, although it's not perfect, it remains the most authoritative summary of what we know. See the IPCC, "Summary for Policymakers," in M. L. Parry, O. F. Canziani, J. P. Palutikof, P. J. van der Linden, and C. E. Hanson (eds.), *Climate Change 2007: Impacts, Adaptation and Vulnerability*. Contribution of Working Group II to the Fourth Assessment Report of the Intergovernmental Panel on Climate Change (Cambridge: Cambridge University Press, 2007), 7–22 (download at http://www.ipcc.ch/pdf/assessment-report/ar4/wg2/ar4-wg2-spm.pdf).

Kurt M. Campbell et al., *Age of Consequences: The Foreign Policy and National Security Implications of Global Climate Change* (Washington, DC: Center for a New American Security, Center for Strategic and International Studies), 2007. This document discusses how climate change is a destabilizing force for societies, and how this may impact national security (download at http://csis.org/files/media/csis/pubs/071105_ageofconsequences.pdf).

ACIA, *Arctic Climate Impact Assessment* (Cambridge: Cambridge University Press, 2005). The Arctic is expected to see the greatest warming and experience some of the most dramatic consequences of climate change. This document discusses in detail the impacts of climate change on the Arctic (download at http://www.acia.uaf.edu).

Ocean acidification is one of the most certain, and underappreciated, impacts of climate change. For a short discussion, see R. A. Kerr, "Ocean Acidification Unprecedented, Unsettling," *Science* 328 (2010): 1500–1501 (doi: 10.1126/science.328.5985.1500).

Here are a few other books I recommend that describe the impacts of climate change. They are all aimed at the general public, so they are easy reads and require no specialized knowledge: E. Kolbert, *Field Notes from a Catastrophe: Man, Nature, and Climate Change* (New York: Bloomsbury USA), 2006; M. Lynas, *Six Degrees: Our Future on a Hotter Planet* (Washington, DC: National Geographic, 2008); J. Diamond, *Collapse: How Societies Choose to Fail or Succeed* (New York: Penguin, 2011).

Problems

1. Your third cousin once removed asks you why we won't be better off in a warmer climate. What do you tell him?
2. Your friend says, "Climate scientists are such alarmists. First they say that floods will become more frequent, and then they say that droughts will become more frequent. Come on, which one is it? They both can't occur!" What do you tell her?
3. As discussed in this chapter, temperatures are not expected to rise uniformly across the globe.
   a) Why is there more warming at high latitudes than the tropics?
   b) Why will land warm more than the ocean?
   c) Why do temperature contrasts (e.g., night vs. day) decrease in a warmer climate?
4. Precipitation
   a) How is precipitation expected to change in a future climate?
   b) Why do changes in the form of precipitation (rain vs. snow) matter?
5. Explain a few ways that climate change impacts public health.
6. Why will it be easier for the United States and Western Europe to deal with climate change than countries in Africa?
7. Explain how climate change affects our national security.
8. a) What do scientists mean when they talk about "abrupt climate change"?
   b) Give an example of an abrupt climate change that has occurred in the past.

Exponential growth

Exponential growth may be the most important term that you've never heard of. It touches many aspects of your life, from the growth of credit card debt and housing prices to governing key processes in biology, physics, economics, and, yes, climate change. In this chapter, we will explore exponential growth and its implications, particularly for climate change.

10.1 What is exponential growth?

First, a definition: *exponential growth* means that the rate of growth is directly proportional to the present size. A good example of exponential growth is the accumulation of money in a savings account. Let's assume that you deposit $100 into a bank account with an *interest rate* of 10% per year.[1] After the first year, you receive interest equal to 10% of the balance of $100, which is $10. This interest raises the balance of the account to $110. After a second year, the interest is 10% of the balance of $110, which is $11. This increases the balance to $121. Table 10.1 shows the growth of the bank account over 101 years.

This growth is exponential because the increase in the bank balance in any year is proportional to the bank balance in that year. In fact, anything growing at "x% per year" is growing exponentially.

The key parameter in exponential growth is the rate of growth, or r. For bank balances, credit cards, or mortgages, r is usually called the interest rate, whereas in other contexts r may have other names (later in the chapter, I will refer to r as the discount rate). Usually, r is expressed in percent per year. Given a growth rate of r percent per year, an initial quantity P will grow by a factor of $1 + r/100$ in 1 year. So, after 1 year, the quantity has grown to $P(1 + r/100)$. For a bank balance of $100 and an interest rate of 10% per year, the bank balance after 1 year is $100(1 + 10/100) = \$100(1.1) = \$110$, the same answer we obtained earlier.

At the end of 2 years, the balance is $P(1 + r/100)(1 + r/100)$. This is simply the balance at the end of the first year, $P(1 + r/100)$, multiplied by another factor of $1 + r/100$ to account for growth during the second year. Thus, the bank balance at the end of the second year is $\$100(1 + 10/100)(1 + 10/100) = \$100(1.1)(1.1) = \$110(1.1)^2 = \$121$.

---

[1] We're assuming here that the interest is compounded annually.

Table 10.1 Calculation of the balance of a bank account		
Year	Interest generated ($)	End-of-year balance ($)
1	10	110
2	11	121
3	12.10	133
4	13.30	146
5	14.60	161
6	16.10	177
7	17.70	195
8	19.50	214
$\vdots$		
100	125,278	1.38 million
101	137,806	1.52 million

Note: The initial investment was $100 (i.e., the balance at the beginning of Year 1 was $100), and the account has an interest rate of 10% per year.

You may well be able to see a pattern here. After n years, an initial investment of P will grow to a final value F:

$$F = P(1 + r/100)^n \qquad (10.1)$$

This is the formula I used to generate the values in Table 10.1.

10.2 The rule of 72

One of the most useful concepts about exponential growth is the idea of a *doubling time* – the length of time that it takes for something growing exponentially to double. From Table 10.1, we see that $100 invested at 10% per year will double in approximately 7 years.

A simple way to estimate the doubling time is to use the *rule of 72*: The doubling time is 72 divided by the growth rate (in percent per year). Using this equation, we see that the doubling time at 10% per year is $72/10 = 7.2$ years, a result that is consistent with Table 10.1.

Using this rule, you can frequently do exponential growth problems with pencil and paper – or even in your head. For example, let's put $100 in the bank at 7.2% interest in Year 2000. What is the balance in Year 2100? The doubling time is $72/7.2 = 10$ years, so that the balance doubles every 10 years. Table 10.2 shows the balance at the end of every decade.

After 100 years the $100 investment has grown to $102,400 – illustrating the power of exponential growth. In equation form, an initial investment of P has grown after n doublings to a final value F:

$$F = P(2^n) \qquad (10.2)$$

Table 10.2 The balance of $100 invested in Year 2000 at an interest rate of 7.2% per year		
Year	No. of doublings	Value ($)
2000	–	100
2010	1	200
2020	2	400
2030	3	800
2040	4	1,600
2050	5	3,200
2060	6	6,400
2070	7	12,800
2080	8	25,600
2090	9	51,200
2100	10	102,400

This is the equation I used to calculate the values in Table 10.2. And in a pinch, I could have done the calculation with just pencil and paper.

We could also have used Equation 10.1 to calculate the balance in 2100: $F = \$100(1 + 7.2/100)^{100} = \$104{,}587$. This is very close to the value calculated for 2100 by use of the doubling time, although the estimates differ slightly. The difference results from the fact that the rule of 72 is approximate, so calculations using it are almost always slightly off. In most cases, though, the rule of 72 is accurate enough.

As another example, imagine investing $100 at an interest rate of 14.4%. How long would you have to leave this investment in the bank to yield $1 trillion ($10^{12}$)? First, let's figure out how many doubling times it would take. Using Equation 10.2, we can write the relevant equation as

$$\$100(2^n) = \$10^{12} \tag{10.3}$$

Solving this equation, we find that $n = 33.2$. In other words, an initial investment of $100, doubled 33.2 times, yields $1 trillion. Now let's calculate how many years that is. At an interest rate of 14.4% per year, the doubling time is $72/14.4 = 5$ years. That means 33.2 doublings is 166 years. The growth of this investment is plotted in Figure 10.1.

It is important to realize that most of the accumulation occurs in the last few doubling periods. In fact, $500 billion dollars, half of the total, is earned in just the last doubling period, and 97% of the $1 trillion is earned in the last five doubling periods (25 years). Thus, the exponential growth is heavily weighted toward the very end of the investment period.

10.3 Limits to exponential growth

Where I live, College Station, Texas, the population growth rate peaked at 11% per year in the 1970s. At that rate, College Station's population was doubling every 72/11

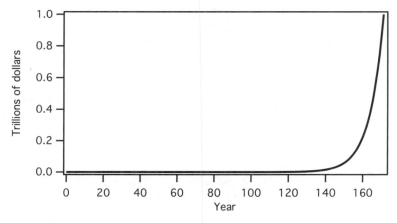

Fig. 10.1 Value of $100 invested at 14.4% interest as a function of years invested.

= 6.5 years, or about 15 doublings per century. That growth rate, sustained over a century, would result in College Station's population increasing by a factor of $2^{15} =$ 33,000. This means that, given a present-day population of 70,000, at that rate the population would reach 2 billion by the end of the 21st century. This is an impossibly high population and means that this growth rate could not possibly be sustained for 100 years. In fact, considering the practical limits of city growth, such high population growth is unsustainable for much more than a decade or two. Thus, it should have been obvious in the 1970s that the population growth rate of College Station would decrease, and it has, to roughly 3% per year today.

A growth rate of 3% per year sounds much more sustainable, and it is, but exponential growth is so stunningly fast that even seemingly low growth rates can be unsustainable for more than a few centuries. For example, at 3% per year, College Station's population would approach 1 billion in about 300 years or so. This is also clearly unattainable in any practical sense, and it says that, over the upcoming centuries, the population growth rate of College Station must decline even further.

This brings us to the first important rule of exponential growth, best expressed by economist Kenneth Boulding:

> Anyone who believes exponential growth can go on forever in a finite world is either a madman or an economist.[2]

A second rule of exponential growth is that, when the end comes, it comes quickly. As an example, let's consider a resource consumed by humans, such as water. Let's assume that in Year 2000, a community is consuming 100 million gallons of water per year. Let's also assume that the amount of water consumed grows at 7.2% per year, meaning that consumption doubles every 10 years. The local reservoir that supplies the water can supply a maximum of 100 billion gallons of water per year. The growth over time of the fraction of the reservoir's capacity that is consumed is plotted in Figure 10.2.

---

[2] Quoted in Deffeyes (2006).

Table 10.3 The amount of water consumed each year by a community, assuming that consumption increases by 7.2% per year		
Year	Gallons per year consumed	Fraction of supply consumed (%)
2000	100 million	0.1
2010	200 million	0.2
2020	400 million	0.4
⋮		⋮
2070	12.5 billion	12.5
2080	25 billion	25
2090	50 billion	50
2100	100 billion	100
2110	200 billion	200
2120	400 billion	400

Note: The middle column shows the water consumed each year by a community. Dividing the rate of consumption by the maximum amount that can be supplied (100 billion gallons/yr) yields the fraction of the supply that is being consumed (right-hand column).

Table 10.3 lists the amount of water consumed every decade, as well as the fraction of the total supply consumed. In Year 2000, the community is using just 0.1% of the total available water. This is such a small amount compared to the size of the supply that the community members don't even consider the idea that they might one day run out of water. However, this fraction doubles every 10 years, and by Year 2070 the rate of consumption has increased more than 100-fold, to 12.5% of the supply. That very year, a small group of activists claim that the world is on the verge of exhausting its water supply. Most citizens dismiss this claim out of hand: The community is only using a small fraction of the total supply, and it had taken 70 years of growth to go from 0.1% to 12.5%. Based on this, the majority argue, it will be centuries before the community hits the limit of the water supply.

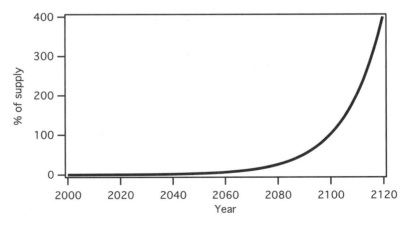

Fig. 10.2 Consumption of water as a fraction (in percent) of total supply, based on the data in Table 10.1.

But the majority is wrong. As Figure 10.2 shows, around Year 2070 something remarkable happens: The consumption of water "turns the corner" and begins heading up rapidly.[3] The reason is that, during exponential growth, the increase during the next doubling period is equal to the increase during all of the previous doubling periods, combined. In other words, the increase in consumption between 2070 and 2080 is equal to the total increase over the entire period before 2070. So even though it took a very long time for the world to reach a consumption rate of 12.5 billion gallons per year, the next doubling period sees an increase in 12.5 billion gallons per year. Moreover, it will only take three more doublings – until 2100 – before consumption is 100 billion gallons per year, or 100% of the supply.

Now let's assume that the community suddenly realizes that it's going to reach its limits of consumption in a few decades. Through Herculean efforts to develop and deploy new technology, the availability of water is increased by a factor of 4 – so the new limit is 400 billion gallons per year.

That's an enormous increase – but it buys you far less time than you might think. Given that the community will be consuming 100 billion gallons in 2100, 200 billion gallons in 2110, and 400 billion gallons in 2120, the new water supply limit is reached in just two doublings or 20 years. The lesson here is that, when things are growing exponentially, resource limits may be much closer than you think.

In reality, of course, important economic factors come into play. As limits to the resource are approached, the price should go up, giving consumers a clear economic signal to change their behavior. This will encourage conservation, as well as technical development of new sources and of substitutes.

Another good and recent example is the subprime housing crisis that hit the United States in 2007. Beginning in 2004, houses in some places began appreciating at 20–30% per year, corresponding to a doubling time of 3 years or so (Figure 10.3). At that rate, a $250,000 house would become a $2 million house in 10 years (approximately three doubling periods). Common sense tells us that's simply impossible – at that rate, houses would rapidly become unaffordable.

Although this conclusion may seem obvious in retrospect, at the time people seemed to believe this exponential growth could go on forever. Homebuyers were willing to pay ever-increasing prices and take out loans they could not possibly afford because they believed that the value of their house would always continue to appreciate. If they got in trouble, they could always sell the house at a profit. Investors believed the same thing, so they were willing to fund the absurd mortgages. Eventually, buyers with insanely large mortgages began defaulting on them, leading to a sudden, rapid collapse of the housing market. Just a few years after houses were appreciating at 30% per year, their value was declining at 20% per year.

Probably the best-known warning of the dangers of exponential growth came about 200 years ago from *Thomas Malthus*. He argued that population grows exponentially whereas food production grows in a linear fashion. Linear growth means that the

---

[3] If the exponential were infinite, then the choice of where the exponential turns the corner is arbitrary and can be selected by carefully choosing the scale. However, when it is exponentially growing consumption of a fixed resource, then the region where the curve turns the corner is real and set by the maximum value of the resource.

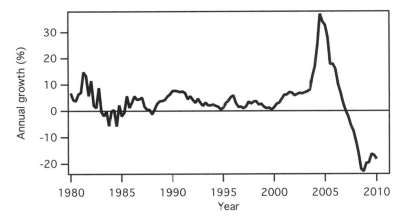

Fig. 10.3 Annual growth in house prices in Nevada (data obtained from the St. Louis Federal Reserve Web site; see http://research.stlouisfed.org/fred2).

increase in each year is a fixed amount: Food production after n years as $F(n) = an + b$, where a and b are constants and n is the number of years. This is quite different from exponential growth, in which the increase is a fixed fraction of the previous year's amount. Mathematically, it is easy to show that exponential growth will always eventually outpace linear growth, and this led Malthus to conclude that an exponentially increasing population would eventually outstrip the world's ability to feed those people, resulting in widespread starvation – and what we now call a Malthusian catastrophe.

Malthus was correct that population grows exponentially. However, in the two centuries since Malthus' prediction, technological developments (e.g., development of fertilizers and pesticides) have allowed food production to increase exponentially along with population. As a result, we have not experienced this Malthusian catastrophe. We are not, however, out of the woods yet. One of the main points of this chapter is that exponential growth cannot continue indefinitely: Both food production and population will eventually cease to grow exponentially. The question is which one plateaus first. If exponential food production growth stops before exponential population growth does, then Malthus's prediction may yet come true. If, however, population growth ceases before food production growth, then Malthus may be forever wrong. Time will tell.

For climate change, we are also interested in population growth because that is one of the main drivers of emissions. We know that population must eventually cease growing – there must be some limit to how many people are living on the planet. Exactly where it levels off, though, is a crucial determinant of our future climate. Predictions of future climate are also dependent on predictions of future consumption. In the recent past, U.S. economic growth has been roughly 3% per year, suggesting that U.S. wealth will double 4 times per century. This rate of growth implies that U.S. citizens will be consuming 16 times as much in 100 years as we are consuming today, consuming 256 times in 200 years, and 4,000 times in 300 years. China is an even more dramatic example – it has been growing at around 10% for more than a decade. This is a doubling time of approximately 7 years, which translates into 14 doublings per century. At this rate of growth, a Chinese citizen in 100 years will be consuming about 16,000 times as much as he or she is today.

Common sense tells us that even the relatively modest economic growth of the United States must cease within a century or two, and the overheated rate of growth of China must significantly decrease within a decade or two. And it's certainly possible that the end of exponential economic growth may occur much sooner.

10.4 Discounting

In this section, I describe the financial concept of discounting. As I will explain here, discounting plays a key role in evaluating policy options for dealing with climate change.

10.4.1 The time value of money

Suppose you know today that you will incur an expense of $25,000 in 15 years. Such eventualities occur frequently in the world of corporate finance and other business areas. How much would you be willing to pay *today* to eliminate that future expense? One way to think about this is to determine how much you would have to invest today in order to have $25,000 in 15 years. We can get the answer by rearranging Equation 10.1:

$$P = F/(1 + r/100)^n \tag{10.4}$$

Here F is the expense, which will be incurred n years, r is the interest rate in percent, and P is the amount you need to invest today. Given an interest rate of 5%, we need to invest about $12,000 today in order to have $25,000 in 15 years. In other words, we can view $12,000 today as being equal to $25,000 in 15 years.

One conclusion you can draw from this is as follows: *Money in the future is worth less to you than money today*. That general conclusion is probably obvious to most people, but this calculation allows us to answer the question of how much less. The parameter r in these projects quantifies the rate that money loses value as it recedes into the future – each year, money loses r percent of its value. In these types of problems, r is frequently referred to as the *discount rate*, and the value today of a future expense or benefit is referred to as the *present value*. The process of calculating the present value of a future cost or benefit is referred to as *discounting*, and it is one of the central concepts in corporate finance.

An example: What would you do?

You can use discounting to help make financial decisions. Imagine you walk into an electronics store, searching for a new television. You select one and are informed that you have two payment options: You can get it "with no money down" and pay $1,100 in 1 year, or you can pay $1,000 today. Which option do you choose?

Note that $1,100 in 1 year has a present value of $1,100/(1 + r/100), where r is the discount rate. If you choose a discount rate of, say, 5%, then the present value is $1,047. This is more than $1,000, meaning that $1,100 in 1 year is more expensive than $1,000 today. You want to pay as little as possible for the television, so you therefore prefer to pay $1,000 today.

For a discount rate of 15%, the present value of $1,100 in 1 year is $956, so in that case $1,000 today is more expensive than $1,100 in 1 year – and you'd therefore prefer to pay $1,100 in 1 year. If you choose a discount rate equal to 10%, then the present values are equal, and you would have no preference about paying $1,000 today or $1,100 in one year.

In the policy debate over climate change, discounting plays a central role in comparing policy options. Much like the television purchase in the example, our choice is between spending money today to reduce emissions of greenhouse gases, thereby reducing the impacts of climate change in 50 to 100 years, or do nothing now and spend more money dealing with the impacts of climate change in a few decades. So, for example, our choice might be to spend $100 billion today or spend $1 trillion in 100 years.

Discounting allows us to express all costs on a common scale, so we can compare these two options. If we assume a discount rate of 3%, then the present value of $1 trillion in 100 years is $10^{12}/(1.03^{100}) = \$52$ billion. This is less than the alternative of spending $100 billion today, so (from a purely financial perspective) we would prefer to pay $1 trillion in 100 years than $100 billion today. This type of analysis, in which you compare the present value of the costs and benefits of various options, is known as *cost–benefit analysis*, and economists frequently use it to provide guidance to policy makers about alternative policy options.

10.4.2 The discount rate

In both the television purchase and climate examples just given, the answer we get is strongly dependent on the discount rate. In the climate example, a discount rate of 3% yields the conclusion that we'd prefer to do nothing now and pay later to address the impacts of climate change. However, if the discount rate were 2%, then the present value of $1 trillion in 100 years is $138 billion, and we'd rather pay $100 billion today to reduce emissions.

So how do we determine the correct discount rate? The discount rate really is a combination of two different judgments. First is what is known as *time discounting*, which is the preference to consume now rather than later. If offered $100 now or $100 in 1 week, just about everyone would choose to get the money now. After all, why would you wait? Experiments show that animals also exhibit this behavior. I know that if I give my dogs the choice of having dinner now or in an hour, they tell me in no uncertain terms that they want to eat *now*. In other words, most people (and dogs) have a positive time discount rate: Goods and services in the future are worth less than the same goods and services are right now.

Over the long time periods covered by the climate problem, however, the time discount rate really represents our preference for us to consume rather than future generations. Given that consumption can be roughly equated to welfare, the time discount rate then expresses how much we value our welfare above the welfare of future generations. From a moral standpoint, most people agree that it is unethical to place a higher value on our own welfare over that of future generations, which implies that the time discount rate should be set to near zero. Nonetheless, it's clear from our society's actions, such as our low rate of savings and our failure to address big problems facing future generations (e.g., climate change, budget deficits), that we do indeed value consumption by our own generation more highly, which implies a positive time discount rate.

The other part of the discount rate is known as *growth discounting*, and it reflects the fact that a fixed amount of money means more to poor people than it does to rich people. For example, if a billionaire is walking down the hall and sees a $5 bill on the floor, would he stop to pick it up? Probably not – if you have 1 billion dollars, another 5 dollars does not do anything to improve your welfare. If you're living in poverty, however, you're most certainly going to stop and pick up the $5 bill – it might be the difference between having dinner that night or not.

In economics jargon, the utility of $5 to the billionaire is much lower than the utility of $5 to the person living in poverty. In the case of climate change, we expect future generations to be richer than we are, just like we're richer than those living 100 years ago. And because future generations are richer, they will be better able to pay costs associated with climate change than we are. This suggests that our generation should consume in order to improve our welfare (particularly those living in poverty today) and push the costs of addressing climate change onto the richer future generations. This preference is expressed in the growth discount rate, which is the rate at which the utility of money – how much $1 means to society – declines with time as the world gets richer.

The discount rate used in present-value calculations is determined by combining the time and growth discount rates. Unfortunately, the choice of both time and growth discount rates is as much of a value judgment as an objective fact, and there are wide disagreements about what discount rate to use. Some economists argue the discount rate should be near zero, whereas others argue for higher values such as 4%. In Chapter 14, I will discuss in more detail the implications of uncertainty in the discount rate on our evaluation of the cost–benefit analysis of the climate change problem.

10.5 Chapter summary

- Exponential growth means that the rate of growth is directly proportional to the present size. Anything growing at "x% per year" is growing exponentially.
- A quantity P growing at r percent per year will grow to $P(1 + r/100)^n$ after n years. The time for a quantity growing exponentially to double is frequently referred to as the doubling time; it is approximately equal to $72/r$, and this shortcut is known as the "rule of 72."

- Exponential growth tends to be end loaded, meaning most of the growth occurs at the end. In particular, 50% of the growth occurs during the last doubling period, and 97% of the growth occurs during the last five doubling periods.
- Exponential growth cannot go on forever, and when it ends, it often ends suddenly.
- Discounting refers to the process of calculating the present value of some future expense or benefit. Such calculations require a discount rate, which is the rate at which money loses value in the future. In general, money in the future is worth less than money today.
- Cost–benefit analyses compare the present values of the costs and benefits of a range of policy options. The best option (from a financial point of view) is the one with the largest present-value net benefit.
- The discount rate is determined by two judgments: the time discount rate, which is our preference for consuming now rather than later, and the growth discount rate, which reflects the fact that future generations are expected to be richer so they can pay a bigger share of the costs. There are vigorous disagreements over what discount rate we should use.

Terms

Cost–benefit analysis
Discounting
Discount rate
Doubling time
Exponential growth
Growth discounting
Interest rate
Malthus
Present value
Rule of 72
Time discounting

Problems

1. You invest $1 at a 10% interest rate for 50 years.
   a) Use Equation 10.1 to calculate how much you have after 50 years.
   b) How many doubling periods does the investment experience?
   c) Use Equation 10.2 to calculate how much you have after 50 years.
2. You invest $50 at a 7% interest rate for 30 years.
   a) Use Equation 10.1 to calculate how much you have after 30 years.
   b) How many doubling periods does the investment experience?
   c) Use Equation 10.2 to calculate how much you have after 30 years.

3. a) How many doubling periods do you have to wait for 1 cent to grow to $100 trillion? (Calculate to the nearest integer.)

   b) At an interest rate of 7%, about how long does it take for that many doublings to occur?

4. Would you rather pay $1 trillion dollars of damages from and adaptation to climate damage in 50 years or pay $20 billion dollars today to reduce emissions and avoid the climate change? Use discount rates of 0%, 2%, 4%, 6%, and 8%.

5. You go into a big-box electronics store to buy a flat-screen television. You have two options: pay $1,400 today or $1,450 in 1 year. Which do you choose? You have to estimate a discount rate to do this. How did you choose your discount rate?

6. Lotteries often give you the option of taking a lump-sum payment now or a fixed amount every year for, say, 25 years. For this question, assume that the lump-sum payment is $3 million and the yearly payments are $250,000 each year for 25 years. The first payment is made immediately, so it is not discounted, and subsequent payments are made every year thereafter.

   a) Find the discount rate where the present value of the 25-year cash stream is equal to the lump-sum payment.

   b) If your discount rate is higher than this, should you take the lump-sum payment or the period payments?

7. In the National Football League draft, a pick in this year's draft is worth a pick in a lower round in a future draft (e.g., you might trade a second-round pick in this year's draft for a first-round pick in next year's draft). Explain how this is consistent with the concept of discounting.

8. a) Imagine you have a dollar bill. If you double it, you have two bills. If you double again, you have four bills. If you double again, you have eight bills, and so on. Given that a bill is 0.1 mm thick, how many doublings do you have to go through before you have a stack that reaches from the Earth to the moon? (The moon is 360,000 km away.)

   b) How many doublings do you need to get a stack that goes halfway to the moon?

   c) How many doublings to you need to get 1% of the way to the moon?

9. a) Consider the choice between paying $10 million today to reduce emissions that cause climate change or $1 billion in 100 years to adapt to a changing climate. What would the discount rate have to be in order for these two choices to be equal?

   b) Using that same discount rate, what would be your preference if the expense was in 50 years instead of 100?

10. You are inside the Houston Astrodome, in the rafters just below the roof, 160 ft above the field. They put a tiny magic drop of water on the pitcher's mound, and the drop starts doubling in volume every minute. After 100 minutes, there is 5 ft of water on the field, and the depth continues to double every minute. How many minutes do you have before the water reaches you?

11

Fundamentals of climate change policy

In the previous chapters of this book, we have seen that 1) the Earth is warming, 2) most of the recent warming is very likely due to human activities, 3) warming over the next century will likely be a few degrees Celsius, and 4) such warming carries with it a risk of serious, perhaps even catastrophic impacts for humans and the rest of the planet's ecosystems.

Given those facts, what shall we do about climate change? Science, it turns out, is just one of several factors needed to answer this question. Deciding what to do also requires information about the options available to us to respond to climate change, and the costs, benefits, and risks of each option. In addition, we must consider not just monetary costs but also the moral trade-offs inherent in each policy. In this chapter, I will outline the various options available to us to address climate change.

Our responses to climate change can be broadly split into three categories: adaptation, mitigation, and geoengineering. *Adaptation* means responding to the negative impacts of climate change. If climate change causes sea-level rise, an adaptive response to this impact would be to build seawalls or relocate communities away from the encroaching sea. *Mitigation* refers to policies that avoid climate change in the first place, thereby preventing impacts such as sea-level rise from occurring. This is accomplished by reducing emissions of greenhouse gases, usually through policies that encourage the transition from fossil fuels to energy sources that do not emit greenhouse gases.

Geoengineering refers to active manipulation of the climate system. Under this approach, our society would continue adding greenhouse gases to the atmosphere, but we would intentionally change some other aspect of the climate in order to cancel the warming effects of the greenhouse gases. For example, one geoengineering approach is to intentionally increase the albedo of the Earth, thereby offsetting the warming of greenhouse gases. In the rest of this chapter, we explore each of these options in detail.

11.1 Adaptation

As temperature, precipitation, sea level, and other components of climate change, we can adapt our way of life to adjust to these changes. Strictly speaking, adaptation does not require a collective decision be made to adapt. When individuals realize they are being harmed by a changing climate, they will initiate action to reduce the harm. If sea-level rise threatens to inundate a community, each individual can make his or her own decision whether to move, or remain and try to live with the higher sea level.

Citizens of a community can also undertake coordinated action. For example, the citizens of the community may decide to band together and build coastal defenses, such as a seawall – an option that is too expensive for any individual.

In certain situations, national and international government assistance to help a community adapt makes sense. For example, if sea-level rise submerges Miami, and the resulting economic disruption hurts the entire U.S. economy, then the U.S. government might be justified in paying for seawalls to prevent that from happening. Or if climate impacts in China threaten to destabilize the entire world economy, international assistance to help the Chinese deal with the impacts may be appropriate. Similarly, national and international government assistance could be viewed as a form of social insurance that allows risk sharing across the national or international community. Adaptation might also be considered a form of aid, whereby richer societies agree to help poorer ones pay for climate impacts for purely distributional reasons.

In addition to direct aid, governments can also implement regulations to encourage citizens to adapt to a changing climate. Regulations promoting water conservation, for example, would help communities adapt to decreased freshwater availability. Governments can also eliminate existing regulations that encourage us to be poorly adapted to the present climate and that increase our vulnerability to climate change. A good example is flood insurance. People love to live near bodies of water, such as the ocean. However, the downside of living near water is that flooding may occasionally destroy their houses. Without flood insurance, many people would not build in flood-prone areas because they could not afford to have their houses destroyed. With flood insurance, however, people can afford to live in flood-prone areas; if their houses are destroyed by flood, the insurance covers the loss. In this way, flood insurance actually encourages people to build where it's going to flood.

A third way government policy can facilitate adaptation is by providing reliable information on the climate changes that will occur, as well as possible responses, including technical assistance. Such information would allow people to take action *before* climate change occurs. Adaptation in advance is always cheaper than adaptation after an event. One reason is that equipment and built structures (roads, buildings, houses, airports, power plants, etc.) tend to have lifetimes of decades. By knowing climate change is coming far in advance, one can design infrastructure and equipment to be resilient to climate change over its entire lifetime. Otherwise, infrastructure that still has useful life left may have to be scrapped, which is an expensive eventuality.

Imagine, for example, that you were building an airport on the coast that was designed to last 100 years. If the seas submerge the airport after, say, 20 years, then the location would have to be abandoned even though it had many useful decades left, and a new airport would have to be built at a higher elevation. This would impose a large and avoidable cost on the community. Thus, you would like to be able to pick a location for the first airport that was not vulnerable to sea-level rise over the useful life the facility – and having good information about sea-level rise would enable the community to make the appropriate choice.

There are several advantages to relying on adaptation as the main response to climate change. First, because many of the worst impacts of climate change will occur in the second half of the 21st century, adaptation allows us to wait for decades before we must start adapting. There are good reasons to do this – for example, it allows us to resolve uncertainty on what will happen or on how to respond. Moreover, by waiting until the impacts become more certain, the adaptation that we do undertake will be more sharply focused on the impacts that occur, without any wasted effort. We also avoid the scenario in which our society spends a lot of money today reducing emissions of greenhouse gases, only to find out later that climate change is just not that serious of a threat. Another reason to wait is that, if the past is any guide, we can expect future generations to be much richer than we are and better able to bear the costs of adaptation (this is the factor that growth discounting accounts for).

Another advantage of adaptation is that it does not require national or international intervention. When the sea level rises, if national governments or international organizations do nothing, then it is left up to local communities or individuals to decide whether to build coastal defenses, move to higher ground, or take their chances. Because of this, those philosophically opposed to government regulation find adaptation to be a less serious infringement on personal liberty than policies that require national and international regulations.

Finally, many of the adaptations necessary to address climate change will simultaneously benefit society in other ways. Decreasing a community's vulnerability to sea-level rise caused by climate change will also decrease the vulnerability to extreme sea-level events caused by hurricanes or other severe storms. Conserving water to address decreased freshwater availability will also decrease the community's vulnerability to droughts or to increased demand caused by population growth. Furthermore, an improved public health infrastructure designed to head off disease outbreaks in a warmer world would also help decrease a society's vulnerability to pandemic flu and other nonclimate public health issues.

But there are also important disadvantages to relying on adaptation as our response to climate change. First, adapting to the impacts of climate change takes significant resources. This means that rich, well-governed places such as the United States or Europe have resources to undertake a serious adaptation effort. Much of the rest of the world, however, is so crushingly poor that its countries can barely provide basic services for their citizens today. They simply do not have additional resources available to them to do things necessary to ameliorate the impacts of climate change. For them, building coastal defenses, developing new freshwater sources, or improving public health infrastructure is simply unaffordable. This also means that pushing adaptation to the local level limits what adaptation can be undertaken, because many adaptive strategies are too expensive for local communities to undertake without help.

This is why the effects of environmental disruption are felt most strongly by the poorest and most vulnerable in any society. This was ably demonstrated when Hurricane Katrina hit New Orleans in 2005. When the storm hit, the wealthier residents of New Orleans, residents with resources such as credit cards and automobiles, simply evacuated. The poorest residents of the city, who lacked resources to evacuate, were stranded in New Orleans and made up the vast majority of those killed. The poor also tended to live in more poorly constructed housing, frequently located in marginal areas, such as swampy or low-lying, flood-prone land. Thus, the damage Katrina inflicted also predominantly impacted the poor.

Once the storm passed, the wealthier residents of New Orleans had the resources to reconstruct their lives – either to return and rebuild or to start anew somewhere else. The poorer residents have had to rely on government assistance to rebuild their lives. Because the U.S. government is rich, it was able to provide assistance such as money and temporary housing that kept Katrina from being a true humanitarian disaster.

We also see connections between vulnerability and wealth in two recent earthquakes. In early 2010, a magnitude-7.0 earthquake ravaged Haiti – it killed more than 200,000 people, made 1 million people homeless, and heavily damaged much of the country's built structures. A few weeks later, a magnitude-8.8 earthquake hit Chile. Although this earthquake was about 500 times stronger, this earthquake killed just a few hundred people. Among other factors, much of the difference in death toll can be attributed to Chile's greater wealth – buildings in Chile were constructed to higher standards and did not collapse during the quake.

The essential lesson is clear: How serious the impacts of climate change are is a function of how wealthy you are. The connection between climate change impacts and wealth means that adaptation fails a fundamental fairness test. The world's rich economies have built their wealth by consuming massive amounts of energy, which means that these economies are responsible for most of the increase in greenhouse gases over the past two centuries. This means that it is the rich countries that are responsible for most of the climate change we are now experiencing. Yet their very richness allows these countries to deal most effectively with the impacts. The poorest countries in the world are responsible for very little of the greenhouse gases in our atmosphere today – which is why they're poor – and it also means that they are least capable of dealing with the impacts.

Because of this, an adaptation-only response is frequently viewed as morally unacceptable because it abandons the poorest people in the world to the impacts of climate change that they did not cause. To the extent that adaptation is necessary, there is general agreement in the international community that rich countries must help the poorer countries of the world deal with the impacts of climate change. Nonetheless, despite the strong agreement over this point, the rich world is doing very little at this point to help developing countries adapt.

Although the rich of the world have many more resources than the poor to deploy against the impacts of climate change, do they have enough? That depends on your definition of "successful" adaptation. If you define successful adaptation as minimal survival of the human species, then humans can successfully adapt to almost any climate change. After all, we are a resilient species and it's hard to believe that any climate change will lead to our total extermination.

A more realistic standard of success is maintenance of our standard of living. The challenge with this metric is that there is no single agreed-upon way to measure it. At its most restrictive, the standard of living can be equated to the amount of goods and services consumed (e.g., GDP). However, standard of living might also include nonmarket activities such as hiking in the same pristine wilderness you did as a child, the satisfaction of seeing the leaves in New England change colors every fall, or appreciating the existence of polar bears, even if you never actually see one. None of these are included in GDP.

Whereas GDP is relatively easy to measure, estimating the value of nonmarket activities is much more difficult. Most of us agree that a world with polar bears is more valuable than a world without, but exactly how much more? The upshot is that a particular strategy for adaptation might be viewed as a success by someone who puts less value on polar bears (or other aspects of the environment) and a failure by someone who puts more. This can lead to starkly different views about our ability to adapt to climate change.

However, there are a few general rules about successful adaptation that we can have high confidence in. First, the more warming we have, the harder it will be to successfully maintain our standard of living – regardless of our definition. In fact, for the most expansive definition of standard of living, one that takes the highest account of the environment, it may be that past climate change combined with committed warming over the next few decades is so large that it is no longer possible for us to avoid significant declines in our standard of living. Second, for a given amount of climate change, how we define success in adaptation plays a key role in determining our ability to successfully adapt. The more expansive the definition, that is, the more value we place on more and more nonmarket aspects of the environment, the harder it will be to successfully adapt.

The bottom line on adaptation: Because of lags in the climate system, as well as lags in the economy, some future climate change is unavoidable. To the extent that this warming cannot be stopped, we must adapt to it. Thus, adaptation must be a part of our response. However, relying *entirely* on adaptation as our response is problematic. Adaptation requires resources, and many of the world's poorest inhabitants have few resources and therefore little ability to adapt. As a result, even low-to-moderate climate change will impose harsh impacts on these people. Many view this as morally unacceptable because the world's poor have contributed little to the problem of climate change.

Rich countries have more resources to address the problem, and for low-to-moderate climate change they may well be able to successfully adapt. But if climate change is at the upper end of the range of predictions discussed in Chapter 8, then even rich countries may not have enough resources to adapt. Thus, relying entirely on adaptation as our only response to climate change is a titanic gamble for the rich that climate change over the next century will not be severe. For the poor, it is no gamble at all – any climate change will impose significant hardships. All of this, of course, depends to some extent on the definition of successful adaptation.

As a result, adaptation-only policies are not seriously considered in the climate policy debate, and there is wide agreement that mitigation must be part of our solution to the problem of climate change.

11.2 Mitigation

Mitigation refers to reductions in emissions of carbon dioxide and other greenhouse gases, thereby preventing the climate from changing in the first place and avoiding the impacts of climate change. Because relying entirely on adaptation is a risky strategy, most policy makers view mitigation as playing a vital role in addressing long-term climate change. There are several approaches that could be used to reduce emissions, and I discuss the range of options available in this section.

Before we get into the details of mitigation, I must remind you that the warming of the climate system lags by a few decades the emissions of greenhouse gases. In addition, there are also economic lags that prevent policies to reduce emissions from having an effect immediately on emissions. Because of these lags, the climate over the next few decades is largely determined by emissions that have already occurred and investments in infrastructure that have already been made. Instead, mitigation efforts made today will significantly reduce climate change during the second half of the 21st century.

The first question is this: How much do we have to reduce emissions? As we will explore in Chapter 13, almost every country in the world has signed the U.N. Framework Convention on Climate Change, a treaty that includes the goal of stabilizing "greenhouse gas concentrations in the atmosphere at a low enough level to prevent dangerous anthropogenic interference with the climate system." This statement is uncontroversial at this level of abstraction. However, turning this into a concrete policy requires us to make a decision about what amount of warming would be considered "dangerous." In the past few years, many experts have made the judgment that warming of more than $2\,^\circ C$ above pre-industrial temperatures would be considered dangerous. Achieving this target would in turn require a reduction in the global emissions by the middle of the 21st century of greenhouse gases by 50–80% below today's emissions levels. We will explore this target more in Chapter 14.

In Chapter 8 we explored the factors that control emissions of greenhouse gases: population, affluence, and technology. Thus, we can recast the problem of reducing emissions into the problem of reducing one or more of these factors until emissions reach a desired value. The first factor is the world's population. With fewer people on the planet consuming goods and services, emissions would certainly decrease. Some societies have implemented policies to actively influence the size of their population. China, for example, adopted a "one-child policy," which limits the number of children a family can have to one, although there are some exemptions. This policy has significantly reduced China's population growth rate, although the total population is still increasing.

However, reducing emissions through population control would require not just a reduction in the rate of population growth, but a large decline in the actual number of people on the planet. Such an effort would conflict with deeply held religious, social, and cultural traditions surrounding reproduction and family size in many countries. As a result, efforts to combat climate change by using policies explicitly targeted at reducing the Earth's population are viewed as politically unachievable, and there are no serious discussions of this approach.

A second option is to reduce the world's consumption of goods and services. If each person consumed less, then the amount of energy consumed, and therefore emissions, would decrease. Like population, solving the climate problem through consumption would require not just stopping growth of consumption, but deep reductions in it. There are several problems with solving climate change this way. First is a political problem – people equate consumption with well-being. That's why all politicians strive for increased consumption (which they call economic growth), and no politician who wants to keep her or his job would agree to a policy that steeply reduces it. Thus, reducing consumption is something that most countries simply will not agree to.

Then there's the 2 billion or so of the world's poorest inhabitants who live in extreme poverty. These people don't have the basic necessities of life – food, clean water, shelter – and meeting those basic needs requires economic development. Such economic development implies increasing consumption. Efforts to limit consumption to address climate change may well mean preventing these people from escaping poverty. Such an outcome is generally viewed as morally unacceptable: We cannot solve climate change on the backs of the world's poorest people.

Thus, like population control, there are no serious efforts to address climate change by reducing the world's level of consumption. The rich world doesn't want to do it, and it is ethically problematic to impose such a policy on the world's poorest people.

If neither population nor consumption has any chance of being reduced, then by process of elimination it is the technology term, also known as the greenhouse-gas intensity, which must be reduced in order to reduce emissions. As I discussed in Chapter 8, the technology term is a measure of how much greenhouse gas is emitted per dollar of GDP. This in turn can be broken into two constituent terms: the energy intensity, a measure of how much energy it takes to generate 1 dollar of GDP (J/\$), and the carbon intensity, a measure of how much greenhouse gas is emitted to generate a unit of energy (CO_2/J). If your memory of this is hazy, you may want to refresh it by reviewing Chapter 8.

Reducing greenhouse-gas intensity therefore requires reducing energy intensity, carbon intensity, or both. Energy intensity is determined to a large extent by the efficiency with which the economy uses energy. Today's society wastes a tremendous amount of energy, and improving our energy efficiency would not only reduce our emissions of carbon dioxide but would have many co-benefits, such as saving us money and reducing air pollution. Because of the co-benefits, energy-efficiency improvements would make sense even if climate change were not a problem.

But can efficiency improvements lead to large enough reductions? To reduce emissions by 50–80% over the next few decades, which is about what's required to stabilize the climate with less than 2 °C of warming, would require reducing emissions by approximately 3% per year. If the world's total GDP (the product of population and affluence) grows by 3% per year, then energy intensity would need to decline by 6% per year or so to achieve the necessary reductions in emissions.

Historically, the energy intensity has decreased at roughly 1% per year. This can likely be maintained, and some improvement may be possible, but most experts believe that rates of decline of energy intensity of 6% per year for several decades are not realistic.

Thus, although improvements in energy efficiency can contribute to emissions reductions, and are something we should be doing now, they are likely only going to play a supporting role in solving the climate problem. We therefore conclude that it is reductions in the carbon intensity term, the amount of carbon dioxide emitted per unit of energy generated, that are required to stabilize the climate.

11.2.1 Technologies to reduce carbon intensity

Reducing carbon intensity is code for switching from conventional combustion of fossil fuels to energy sources that do not release greenhouse gases – often referred to as *carbon-free* or *climate-safe* energy sources. These include nuclear energy, carbon capture and sequestration, and energy sources known as *renewable energy*, because these energy sources are not depleted when utilized: primarily hydroelectric, solar, wind, and biomass energy.

Solar energy is one of the most frequently discussed renewable energy sources. There are actually two different ways to generate energy from sunlight: *solar photo-voltaic* or *solar thermal* methods. Photovoltaic energy is the most common form of solar energy, and you can see it in operation in the form of solar panels located on houses or buildings (and also on satellites and the Space Station). It takes advantage of the fact that, when exposed to light, certain materials such as silicon produce electricity. Solar thermal energy, in contrast, uses mirrors to concentrate sunlight on a working fluid (such as an oil, molten salt, or pressurized steam), heating it to several hundred degrees Celsius. This hot fluid is then used to boil water and drive a turbine, which in turn drives a generator that produces electricity.

Solar energy is in many respects the Holy Grail of renewable energy. As we calculated in Chapter 4, the amount of solar energy falling on the planet is staggering – more than 100,000 TW. This is an enormous amount of energy compared to the amount humans consume, which is about 15 TW of primary power. There are, however, problems with the large-scale adoption of solar energy. One is intermittency – the Sun shines only during daytime and when not obscured by clouds. Thus, solar energy may require some additional mechanisms to ensure the reliable 24-hour availability of power required by consumers. Another is the area required to generate solar energy. Taking into account the intermittency and other efficiency issues, we find that solar energy can supply power at a level of approximately 10–20 W/m^2. To satisfy all human energy needs would therefore require roughly 1 million km^2 to be covered with solar energy collectors, corresponding to 0.2% of the Earth's surface. Although this is a large area, it is comparable to the total area covered by cities, so there's no reason to believe that it is impossible for humans to construct the number of collectors needed for that much solar energy.

Another frequently mentioned renewable energy source is wind. This is a mature technology – the Dutch have been using wind energy for hundreds of years to do useful work, such as pumping water. Today's electricity-generating windmills, often referred to as wind turbines, are quite a bit larger and more sophisticated. The largest ones are 130 m tall, the same as a 40-story building, with 125-m blades. A single one of these wind turbines can generate as much as 6 MW of power, which is approximately 1% of the power produced by a standard-size fossil fuel power plant.

Wind also has the problem of intermittency. The wind does not blow everywhere nor does it blow all the time, so, on average, the 6-MW wind turbine actually produces roughly one third of its maximum value. Taking into account the intermittency of wind, as well as the fact that windmills must be spaced apart so that they do not interfere with each other, we find that a wind farm generates power at a level of approximately 2 W/m^2. Therefore, to satisfy human energy requirements would require covering approximately 1.5% of the Earth's surface area with wind farms containing a total of a few million windmills. It should be noted that putting up windmills does not preclude using the land simultaneously for other activities, such as agriculture.

Although it would undoubtedly be an enormous undertaking to construct the number of collectors and windmills needed to produce that much solar or wind power, human industry has produced amazing feats in the past. During World War II, for example, the countries of the world produced hundreds of thousands of airplanes, tens of thousands of tanks, thousands of ships, and other paraphernalia of war. Compared to that, putting up a few million windmills doesn't sound so hard.

Wind and solar energy have been growing rapidly over the recent past and are emerging as important contributors to our energy supply. However, these sources remain more expensive than electricity from fossil fuels, and the intermittency problem has yet to be solved. Thus, we are not yet on the verge of a wholesale transition of our energy supply to these renewable sources.

Biomass energy is another renewable option; it refers to the process of growing crops and then burning them to yield energy. Because the carbon dioxide released from burning biomass was absorbed from the atmosphere during the growth of the plant, there is no net increase in carbon dioxide in the atmosphere. It is an intuitively attractive energy source, but there are several issues that must be considered. First, the rate of photosynthesis limits the power generated by biomass to roughly 0.6 W/m^2 of farmed land. Thus, to generate 15 TW would require that 15% or so of the land surface be devoted to growing biomass for energy – comparable to the area presently under cultivation today.

This requirement is problematic for several reasons. The first one is obvious: That's a lot of land. We know from experience that much of the additional land will come from clearing forest. This deforestation releases carbon dioxide into the atmosphere, and it causes a host of other local environmental impacts such as loss of native biodiversity and ecosystem degradation. The second reason is that the farming methods used to grow the biomass have to be carefully considered. Production of fertilizer, for example, requires large inputs of energy, mainly from fossil fuels. If fertilizer is used in the growth of the biomass, then it might take as much fossil fuel energy to grow the biomass as is saved by burning the biomass to produce energy.

Finally, it is becoming clear that using food, such as corn, as feedstock for biomass energy severely stresses the food supply. The increased competition for food leads to rising food prices, an impact that falls disproportionately on the poor. The hope is that a technological breakthrough will allow us to produce energy from waste biomass that does not have other uses, such as the waste from corn processing (e.g., corn stalk, corn cobs) or cellulosic biomass such as switch grass. Many scientists are currently working on methods to produce biomass energy from these waste sources.

Despite these difficulties, biomass, particularly in the form of corn-based ethanol, already provides a few percent of U.S. motor fuels and is slated by act of Congress to become a major source of automotive fuel in the U.S. over the next decade or so. For consumers, a transition to ethanol will be relatively seamless. Automobiles that run on gasoline can be modified to use ethanol with a few inexpensive and minor modifications to the engine, and some cars already sold are already capable of burning ethanol (so-called flex-fuel vehicles). The gas stations, refineries, delivery trucks, and all of the businesses that provide gasoline can be adapted to ethanol without major technological problems.

In the end, biomass energy is a promising source of renewable energy. Nonetheless, biomass energy systems must be carefully constructed from end to end to ensure that carbon emissions are actually reduced (e.g., carefully evaluating how much fertilizer is used, what land it is grown on). In addition, new technologies that allow biomass energy to be extracted from nonfood biomass must also be developed. Thus, unlike wind and solar energy production, biomass energy production may not yet be ready for large-scale adoption.

Hydroelectric energy is the most widespread renewable energy source in the world today, providing 16% of the world's electricity. Despite the many advantages of this energy source, it seems unlikely that this power source can be greatly increased. Many of the world's big rivers are already dammed, and new dams often cause local environmental problems and therefore generate significant opposition from those individuals living in the area.

One of the most contentious options for reducing greenhouse-gas emissions is nuclear energy. Currently, nuclear reactors generate nearly 16% of the world's electricity. Although nuclear is not actually a renewable energy source, with the technology to recycle and reprocess spent nuclear fuel, there are centuries worth of uranium in the ground, even assuming a massive expansion of the world's nuclear generation capacity. Nuclear is a mature technology, so there are no questions about its technical feasibility.

Opponents of nuclear energy make several arguments against this form of energy. The first is reactor safety, a problem dramatically demonstrated by the 1986 meltdown of a reactor at Chernobyl, during which errors by the operators caused an explosion and fire in a nuclear reactor. The subsequent release of radioactivity to the atmosphere resulted in an environmental disaster in the region around the reactor as well as radioactive fallout across much of Europe. In addition, nuclear power plants present attractive targets to terrorists, and the prospect of an attack large enough to breach the reactor core and release its radioactive contents to the atmosphere is truly scary.[1] It should be noted, however, that modern Western reactors are already designed to withstand the human error that caused Chernobyl and most conceivable terrorist attacks. However, as the recent meltdown of the nuclear reactors in Fukushima, Japan after the earthquake and tsunami of 2011 showed, it is clear that nuclear power brings with it a set of risks that other forms of generating electricity do not.

Another problem is nuclear waste, which is what comes out of the reactor after the nuclear fuel is burned. This waste is extraordinarily radioactive, and it must be safely isolated for many thousands of years. If it were released accidentally or intentionally released in a so-called dirty bomb, the resulting harm in both human cost and ecological damage could be severe. One way to reduce the quantity of waste is to reprocess the fuel, in which usable isotopes of plutonium and uranium are removed and converted back into fuel for another trip through the reactor. Even with reprocessing, though, some waste must be stored for a very long time – and most people just don't want the waste to be stored near them.

---

[1] This is why this scenario is featured in just about every episode of the television show *24*.

This leads us to the problem of proliferation. As reactor fuel is mined, enriched, and reprocessed, there exists the possibility that small amounts of highly enriched,[2] bomb-grade uranium or plutonium could be acquired with the intent of building a nuclear bomb. This could occur by theft from a legitimate nuclear program, or it could be the explicit goal of a rogue state's nuclear program. The net result would be a nuclear weapon in the hands of terrorists or unstable rogue nations, which would present a significant security threat to the rest of the world. This is why the United States and many other countries are so opposed to Iran's developing a nuclear energy program.

Finally, there's the cost. Although nuclear power plants are relatively cheap to run, they are extraordinarily expensive to build. This is one of the primary reasons that no new nuclear power plants have been built in the United States since the 1970s. It may be that nuclear energy cannot be widely deployed without massive government intervention in the construction finances.

A final option to generate energy without emitting carbon dioxide to the atmosphere is known as *carbon capture and storage*, also known by its initials CCS or *carbon sequestration*. This refers to a process by which fossil fuel is burned in such a way that the carbon dioxide generated is not vented to the atmosphere. Rather, the carbon dioxide is captured and placed in long-term storage. Clearly, CCS is not a renewable energy source.

CCS is almost always used in combination with coal combustion, because coal is abundant and produces large amounts of carbon dioxide per joule of energy produced. An example of a CCS method is to expose the coal to steam and carefully controlled amounts of air or oxygen under high temperatures and pressures. Under these conditions, atoms in coal break apart and react with the water vapor, producing a mixture of hydrogen, carbon dioxide, and several other gases. The carbon dioxide is captured, while the other gases are burned in order to generate electricity.

Once captured, the carbon dioxide must be stored. The most likely place to put the carbon dioxide is to inject it deep underground into porous sedimentary rocks, which are distributed widely around the world. Particularly promising sites include depleted oil and gas fields, unminable coal beds, or deep saline formations. This process is technically feasible and would use many of the same technologies that have been developed by the oil and gas industry to enhance the recovery of oil from aging fields. The capacity of these rocks is large enough that they could conceivably hold all of the carbon emitted by human activities.

Using available technology, approximately 85–95% of the carbon dioxide produced can be captured. This comes at a price, however. A power plant equipped with a CCS system would need to divert 10–40% of the energy generated into capturing and storing the carbon. Thus, adding CCS would add 10–40% to the price of energy, or something like 1 to 5 cents per kilowatt-hour of electricity (onto a typical price in the United States of 5 to 15 cents per kilowatt-hour).

The world has copious reserves of coal, and in our never-ending quest for power, it seems likely that this coal will eventually be burned. CCS may be the best way

---

[2] Bomb-grade uranium or plutonium is enriched in the isotopes that are best able to sustain nuclear chain reactions, e.g., uranium-235.

to simultaneously burn this coal while avoiding climate change. However, although CCS is a promising technology, it remains unproven because no large-scale CCS power plant has ever been built.

11.2.2 Policies to reduce carbon emissions

So the basic question of how to mitigate climate change is really a question of how to encourage the world to switch away from fossil fuels to carbon-free energy sources, such as wind, solar, nuclear, and CCS. Whatever policy we adopt must be able to do this at a sufficiently low cost, with sufficiently little social disruption, and without working at cross purposes with other societal goals, such as reducing world poverty and conserving biodiversity, that it can maintain political support over the decades required to address the problem.

Let's begin by addressing the question of why we need regulations to reduce greenhouse-gas emissions in the first place. If people value a stable climate, then won't the free market take consumers' interests into account and reduce greenhouse-gas emissions without any government intervention? There is, after all, abundant evidence that the free market is efficient at allocating resources and producing socially beneficial outcomes (although, as the occasional economic meltdown shows, it is not perfect). As U.S. Senator Chuck Hagel said in an interview with Web site grist.org,[3]

> I have always believed that the marketplace does work. It works because it's based on one fundamental dynamic, which is self-interest of an individual, a company, or a country. The marketplace fosters competition and always trends toward producing a better, cheaper product, which means it is a driver of efficiency. It's in the interests of everyone here to make a cheaper product that's less energy intensive. It cleans up the environment, which has economic advantages too.

So if switching to carbon-free energy is something we want to do, wouldn't the free market take care of this all by itself? The answer to that question is no – emissions reductions sufficient to stabilize the climate will not occur by themselves.

There is, of course, some truth to Senator Hagel's argument. Because consumers have to pay for electricity, consumers have an incentive to reduce how much electricity they consume. This applies pressure to manufacturers to design, build, and sell equipment that uses as little energy as possible. As a result of this market pressure, just about every piece of equipment that you buy today is more efficient than the comparable piece of equipment that was available a few decades ago. That is one of the primary factors behind the world economy's long-term decrease in energy intensity that was discussed in Chapter 8.

But it's also important to remember that increases in energy efficiency are, by themselves, insufficient to solve the climate change problem. Solving the climate problem requires a large-scale shift toward carbon-free energy that is not currently occurring. Moreover, there's a good reason it's not now occurring – and may never occur without government intervention.

---

[3] The article is available online (see http://www.grist.org/article/hagel/).

To understand why, consider the following scenario. Imagine you own a company that produces widgets. Let's assume that it costs your company $1 to manufacture a widget, and during the manufacture process, some carbon dioxide is released into the atmosphere. This carbon dioxide will cause some climate change, which causes damage to the global environment valued at 10 cents. Thus, the total cost of manufacturing a widget using this process is $1.10. However – and this is important – the widget manufacturer only pays $1; the costs of climate impacts associated with the widget, 10 cents, are borne by everyone in the world. At the same time, the benefit of producing the widget goes entirely to the manufacturer.

Economists call the costs of climate change imposed on the rest of the world by the widget manufacturer an *externality*. In general, an externality occurs when someone takes an action, and this action imposes involuntary costs on others. Emitting carbon dioxide is a classic externality because it leads to global climate change and therefore imposes costs on everyone in the world.

To understand the economic implications of an externality, imagine that someone invents a new process for manufacturing widgets, in which it costs $1.05 to produce a widget but no carbon dioxide is emitted to the atmosphere. In this case, there are no climate impacts, so the total cost is also $1.05, and this cost is entirely borne by the manufacturer. Because total cost of building a widget under the new process is 5 cents cheaper than the total cost for the older process, it would be beneficial to society for widget manufacturer to switch to this new process.

But the widget manufacturer will not switch. The reason is that the widget manufacturer is only paying $1 per widget under the older method, with the rest of the costs being borne by society. Under the new process, the manufacturer pays the entire cost of $1.05. So even though this new process is cheaper to society as a whole, to the widget manufacturer, this new process is more expensive. Thus, because of the externality, the economically preferred outcome does not occur, resulting in what is frequently referred to as a *market failure*.

Externalities and the associated market failures occur frequently in environmental problems in which some profitable economic activity degrades a common asset such as the atmosphere, ocean, or a river – and the costs of that degradation are paid for by everyone. When people can exploit and degrade some common assets for free, then the result is that these assets tend to be overutilized – resulting in a situation known as the *tragedy of the commons*. This is the fundamental economic explanation for why overfishing is depleting stocks of fish in the oceans, logging is destroying the rainforests, and greenhouse-gas emissions are changing the climate.

In the case of greenhouse gases, emitters exploit the atmosphere by dumping carbon dioxide into it and derive the entire economic benefit of polluting, while the costs are paid by everyone else. If a company decides to reduce emissions, in contrast, then the entire cost of reducing emissions is borne by the emitter – while the benefits are spread throughout the society. In such a case, the rational behavior of each emitter is to exploit the common resource and dump as much greenhouse gas into the atmosphere as is necessary to maximize profit. There is no incentive to reduce emissions.

Thus, the root economic cause of the climate change problem is that it is free for emitters to dump greenhouse gases into the atmosphere. Therefore, to solve climate change, most economists argue that the costs associated with the emission of greenhouse gases must be shifted back onto the emitter. In the parlance of economics, we need to internalize the externality. This is also frequently described by the principle of "polluter pays," meaning that emitters should be held accountable for the damage they cause.

If the widget manufacturer had to pay 10 cents in order to emit the carbon dioxide associated with the production of one widget under the old manufacturing technology, then the manufacturer would be paying the entire cost of manufacturing the widget. In that case, it would then be cheaper for the manufacturer to switch to the new method, which costs $1.05 to produce a widget with no emissions of greenhouse gases. This benefits both the manufacturer and society.

The bottom line on mitigation: Most experts view mitigation efforts as an absolute necessity. Without any mitigation efforts, it is likely that warming over the next century will exceed the threshold that most experts judge to be dangerous: 2 °C warming above pre-industrial temperatures. To avoid this amount of warming, emissions have to be reduced below present levels by 50–80% by the middle of this century.

Although there are many ways to reduce emissions, the only politically acceptable way (at present, at least) appears to be through technology. Improvements in energy efficiency are important – and make sense even if climate change is not an important problem – but such changes cannot by themselves produce the deep reductions in emissions required. Rather, changes in energy-generation technology are going to be required. This means switching to technologies that generate energy without emitting greenhouse gases, which include solar, wind, biomass, nuclear, and carbon capture and storage.

To encourage the adoption of new technologies, most proposed mitigation policies put a price on emissions. Right now, emitting carbon dioxide to the atmosphere is free, and the costs of climate change are imposed on everyone in the world. In this situation, there is no incentive for the emitter to reduce emissions of greenhouse gases. Putting a price on emissions solves this by making emitters pay the full cost of their emissions, thereby giving them the appropriate incentive to adopt climate-safe energy technology.

11.3 Geoengineering

A last solution to the climate change problem is known as geoengineering, which refers to actively manipulating the climate system in order to prevent the climate from changing – or even to reverse climate changes that have already occurred. Geoengineering efforts can be roughly divided into two categories: *solar radiation management* and *carbon-cycle engineering*.

Solar radiation management refers to efforts to engineer a reduction in the amount of solar energy absorbed by the Earth, $E_{in} = S (1 - \alpha)/4$, where S is the solar constant and α is the albedo (if you don't remember this, you may want to review Chapter 4). For example, some people have proposed placing a "sun shade" in-line between the Earth and Sun, which would block some of the sunlight reaching the

Earth. This would reduce the solar constant S, which would then lead to cooling of the planet.

Most other solar radiation management schemes cool the Earth by increasing the albedo. Probably the most frequently discussed way to do this is to inject sulfur dioxide (SO_2) into the stratosphere. Once in the stratosphere, this gas reacts with water vapor to form what are known as aerosols – liquid droplets that are so small that they have negligible fall speed. These aerosols reflect sunlight back to space, thereby increasing the albedo of the Earth and leading to cooling (aerosol cooling of the atmosphere was also discussed in Chapter 6). Injection of sulfur into the stratosphere is the same mechanism by which volcanoes cool the planet.

Another option is to increase the reflectivity of clouds. It turns out that the size of the cloud droplets determines how white a cloud is, with smaller cloud droplets making a cloud whiter or more reflective. This is the same reason that powdered sugar, which is made up of small particles, appears whiter than table sugar. Thus, if we could somehow make the particles in clouds smaller, clouds would become more reflective and raise the albedo of the Earth.

One frequently suggested way to do this is to release what are known as cloud condensation nuclei into the cloud. These nuclei serve as seeds that cloud droplets form around. By adding them to clouds, we increase the total number of droplets in the cloud. Because the total water contained in a cloud is basically fixed, this makes the cloud contain more, but smaller, particles. An example of this effect can be seen in ship tracks in the clouds, which were discussed in Chapter 6. As ships steam across the ocean, the exhaust from their diesel engines contains fine particulates that can serve as cloud condensation nuclei. These aerosols are transported by the winds into low-level clouds and brighten them.

There have also been suggestions that houses be constructed with white roofs instead of black roofs in order to increase the albedo. Although this would certainly reduce solar radiation absorbed by the planet, the area covered by roofs is small enough that it would not have a huge impact on the global climate (although it might cool a local urban climate).

The physics supporting these suggestions is robust, and we have high confidence that if any of these schemes were carried out at sufficiently large scale, the planet would indeed cool. There are, however, important disadvantages with these approaches. The first is that solar radiation management schemes focus on temperature, but temperature increases are only one of many impacts associated with climate change – and perhaps not even the most important. We know, for example, that some of the carbon dioxide released into the atmosphere ends up in the ocean, resulting in ocean acidification. Solar radiation management schemes do nothing to address this impact.

Moreover, solar radiation management schemes may create other problems. The 1991 eruption of Mount Pinatubo, for example, led to substantial changes in global precipitation patterns and an increase in the incidence of drought in some regions. We can therefore expect that reducing the amount of solar radiation reaching the Earth, in addition to cooling the planet, would also lead to changes in the amount and distribution of global precipitation. Whether these changes would be better or worse than climate change is a question whose answer is not clear.

And there are important political problems with this approach. Imagine that a few rich countries in the world (e.g., the United States and Europe) got together to inject sulfur into the stratosphere to cool the planet. Then, China or India experienced a severe drought. Whether the sulfur injection caused that drought or not, the countries affected might well believe that it did. It is easy to imagine that this would lead to a great amount of political tension, possibly even the abandonment of the geo-engineering effort. In the worst case, geoengineering by a group of countries might be considered an act of war by another group of countries that suffer some type of weather-related injury at the same time.

In addition, as long as our society is increasing the amount of greenhouse gas in the atmosphere, geoengineering efforts must continually be strengthened in order to provide an ever-increasing cooling influence to keep the climate stable. So if we were, for example, injecting sulfur into the stratosphere in order to enhance the planet's albedo, then we would have to be injecting ever greater amounts of sulfur over time in order to offset a continually increasing abundance of atmospheric greenhouse gases.

Furthermore, if we ever stopped injecting sulfur into the stratosphere, it would take only a few years for the stratosphere to clear (about the same length of time as it does for the climate to recover after a volcano), during which time the Earth's albedo would be rapidly decreasing. This would cause the Earth's temperature to rapidly rise. Thus, once you start geoengineering, it is difficult to stop, particularly if your geoengineering program is canceling out a significant amount of warming.

The second category, carbon-cycle engineering, involves attempts to modify the carbon cycle so that carbon dioxide is more rapidly removed from the atmosphere. As we learned in Chapter 8, the lifetime of a perturbation of atmospheric carbon dioxide is typically a few centuries, and this is the reason carbon dioxide is such a pernicious greenhouse gas. If the lifetime of carbon dioxide could be reduced, then it would reduce the amount of carbon dioxide in the atmosphere – thereby reducing the amount of climate change.

Planting trees is an example of carbon-cycle engineering. As the trees grow, they suck carbon dioxide out of the air and sequester it in wood. Another scheme is to add iron to the ocean. Iron is thought to be a limiting nutrient there, so the addition of iron will stimulate the growth of phytoplankton. As the phytoplankton grow, carbon dioxide will be drawn out of the atmosphere and into the ocean. The phytoplankton is then consumed by larger organisms, and subsequent biological activity creates a flux of dead organisms and fecal matter from surface waters into the deep ocean. This net transport of carbon from the upper layers into the deep ocean is known as the *biological carbon pump*. Thus, adding iron to the ocean has the net effect of drawing carbon dioxide out of the atmosphere and transporting it to the deep ocean.

Another option is to remove carbon dioxide from the air chemically, which is often referred to as *air capture*. This is like CCS, but CCS removes carbon dioxide from the hot exhaust gas of a power plant whereas air capture removes carbon from the free atmosphere. This is an attractive option, but the amount of energy required with the use of today's technology is staggering, and this severely limits our ability to deploy this technology at scales large enough to remove significant quantities of carbon dioxide from the atmosphere.

In general, carbon-cycle engineering is attractive because, unlike solar radiation management, it does not focus just on temperature. If we balance emissions of carbon dioxide from human activities with removal through carbon-cycle engineering, then we would truly stabilize the climate – not just temperature, but other aspects, such as precipitation. In fact, a sufficiently aggressive program could lead to a *reduction* in atmospheric carbon dioxide (not just a stabilization), which would eventually undo most of the effects of climate change (but not all – some changes, such as extinction of species, are completely irreversible, whereas others, such as loss of the world's largest ice sheets, are effectively irreversible on any time scale that we care about).

There are, of course, some problems with the various approaches to carbon-cycle management. First, real engineering of the carbon cycle, such as adding iron to the ocean, is risky. Because of significant uncertainties in our knowledge of how carbon cycles in the ocean, there is a possibility that that these schemes will not work. In other words, we might add iron to the ocean only to find out that, because of unanticipated physics or biology, no extra carbon dioxide was removed from the atmosphere. And adding iron to the ocean might have unforeseen and serious impacts on ocean ecosystems. Thus, in trying to address climate change, we may cause an entirely new environmental problem – and not even solve the problem we were intending to address.

Geoengineering has several general qualities that make it an attractive policy response to climate change. First, and possibly most importantly, it attacks the climate change problem through technology – but does not seek to limit emissions in any way. That allows us to continue doing exactly what we're doing now: burning fossil fuels and consuming energy as fast as we possibly can. Second, geoengineering is a relatively rapid response. Many schemes can be implemented in a decade or so. This contrasts favorably with mitigation, for which emissions reductions must begin now in order to head off climate changes that will occur in 50 to 100 years. Thus, like adaptation, geoengineering is often thought of as something that we will do in the future, if climate change turns out to be bad.

Geoengineering also avoids many of the moral problems of adaptation. A successful geoengineering effort will help the entire planet, not just the rich countries. Moreover, the cost estimates are often cheap enough that one or a few rich countries can shoulder the burden. However, it also raises moral problems of its own: Should we be engineering the climate? Who decides what the ideal temperature is? What happens if geoengineering efforts harm one particular region while benefiting all others? These are all difficult questions that have to be resolved before the world uses geoengineering to address climate change.

The bottom line on geoengineering: Geoengineering is an appealing but risky approach to dealing with climate change. Although it may work, the risk exists that geoengineering may lead to unintended consequences that leave the world worse off. In addition, it may not address all impacts of climate change and may only be viable for a relatively short period of time (less than a century). Thus, in a world where climate change is handled responsibly through mitigation and adaptation, it is unlikely that geoengineering would be needed.

Nevertheless, it is easy to imagine a future where the world makes no progress in reducing emissions. If, by the middle of the 21st century, emissions are large and climate change is out of control, geoengineering may represent the last hope for avoiding truly disastrous climate change. It is in this type of scenario that the deployment of geoengineering is a reasonable strategy. But even here, geoengineering is unlikely to be the final solution. Rather, it would be a way to buy time to let emergency mitigation efforts catch up and begin significantly reducing emissions.

11.4 Chapter summary

- Responses to climate change can be roughly divided into three categories: adaptation, mitigation, and geoengineering.
- Adaptation means learning to live with climate change; mitigation refers to reducing emissions of greenhouse gases, thereby preventing the climate from changing in the first place; and geoengineering refers to active manipulation of the climate system in order to engineer a cooler climate.
- Because not all climate change can be prevented, we will have to adapt to some climate change. However, there are both moral and practical problems with relying on adaptation as our only response.
- Because of the problems with an adaptation-only response to climate change, most experts and world leaders view mitigation as a necessary component of any plan to address climate change. This will be accomplished mainly by transitioning from fossil fuels to energy sources that do not emit greenhouse gases. A key part of policies designed to encourage transition to climate-safe energy is to put a price on emissions of greenhouse gases, which are presently free.
- Geoengineering refers to active manipulation of the climate system. Under this approach, our society would continue adding greenhouse gases to the atmosphere, but we would intentionally change some other aspect of the climate in order to cancel the warming effects of the greenhouse gases. Geoengineering strategies can be broken into two categories – solar radiation management and carbon cycle engineering. Because of potential problems with geoengineering, it is generally considered a last-ditch approach, only to be used if adaptation and mitigation approaches fail.

Additional reading

Much has been written about our policy response to climate change in the form of books, reports, and scholarly publications. Many are available online and you can find them by searching the Internet. Here are a few of particular note.

The IPCC's working group II discusses impacts and adaptation and, while it's not perfect, it remains one of the most authoritative summary of what we know. See the IPCC, "Summary for Policymakers," in M. L. Parry, O. F. Canziani, J. P. Palutikof, P. J. van der Linden, and C. E. Hanson (eds.), *Climate Change 2007: Impacts, Adaptation and Vulnerability*. Contribution of Working Group

II to the Fourth Assessment Report of the Intergovernmental Panel on Climate Change (Cambridge: Cambridge University Press, 2007), 7–22 (download at http://www.ipcc.ch/pdf/assessment-report/ar4/wg2/ar4-wg2-spm.pdf).

The IPCC's Working Group III focuses on mitigation of climate change. See the IPCC, "Summary for Policymakers," in B. Metz, O. R. Davidson, P. R. Bosch, R. Dave, and L. A. Meyer (eds.), *Climate Change 2007: Mitigation*. Contribution of Working Group III to the Fourth Assessment Report of the Intergovernmental Panel on Climate Change (Cambridge: Cambridge University Press, 2007; download at http://www.ipcc.ch/pdf/assessment-report/ar4/wg3/ar4-wg3-spm.pdf).

D. J. C. MacKay, *Sustainable Energy – Without the Hot Air*, (Cambridge: UIT Cambridge), 2009. This is a sober and quantitative analysis of how hard it will be to replace fossil fuel energy with carbon-safe energy sources. The short answer is that it won't be easy (download the book at http://www.withouthotair.com).

E. Kintisch, *Hack the Planet: Science's Best Hope – or Worst Nightmare – for Averting Climate Catastrophe* (New York: Wiley, 2010). This is one of several good books about the good and bad of geoengineering.

Terms

Adaptation
Air capture
Biological carbon pump
Carbon capture and storage
Carbon-cycle engineering
Carbon-free–climate-safe energy sources
Carbon sequestration
Externality
Geoengineering
Market failure
Mitigation
Renewable energy
Solar photovoltaic
Solar radiation management
Solar thermal
Tragedy of the commons

Problems

1. Our responses to climate change can be put into three general categories. List the categories. For each category, give one example of an action that would fall into that category.

2. a) What are carbon-free energy sources? List the ones discussed in the book.
   b) Is carbon-free energy the same as renewable energy?
   c) Is nuclear energy carbon free? Is it renewable?
3. Why do economists generally believe that the free market will not solve the climate problem by itself?
4. What is an externality?
5. Do some research and find an example of "the tragedy of the commons." How is the global warming problem an example?
6. a) Your friend says that "we should rely entirely on adaptation as our response to climate change." Is this a good idea?
   b) I argued here that adaptation must be at least part of our response. Why?
7. a) Explain one way we can "geoengineer" a higher planetary albedo.
   b) Explain one way we can "geoengineer" a reduction in carbon dioxide.
8. a) In this chapter, we divided geoengineering approaches into two categories. What are they?
   b) What are the advantages and disadvantages of geoengineering?
   c) I argued that that geoengineering should be used in what circumstances?
9. a) Imagine a credit card whose bill was divided up and sent to everyone in the United States (i.e., it something was purchased on this card, every person in the United States would get a bill for 1/300,000,000th of your total cost). Would the average person spend freely with this credit card? Or would they be as thrifty as they would if they had to pay the entire bill?
   b) Now imagine that every person in the United States has a credit card like this. What do you think is going to happen?
   c) How is this situation related to the climate change problem?
10. In this chapter, we explored the terms of the IPAT relation and concluded that reducing emissions could really be achieved only through reduction of one term. Which term is it and why is that our only real option?
11. The technology term in the IPAT relation can be further divided into two terms.
    a) For each term, give an example of a technological switch that reduces it.
    b) One of these terms is the key to deep reductions in emissions. Which one is it? What kinds of changes are required to make such deep reductions?
12. Why do mitigation policies have little ability to influence the climate over the first half of the 21st century?

12 Mitigation policies

Chapter 11 discussed the three options we have to address climate change: adaptation, mitigation, and geoengineering. Adaptation will, by necessity, be an important part of our response to climate change. However, relying entirely on adaptation as our only response to climate change is fraught with problems. Geoengineering is another possibility, but one that few people think should be used now. Rather, it is the last resort – like an airbag in a car – that you turn to if other approaches to address climate change fail.

The other option is mitigation – the reduction of greenhouse gases emissions so as to avoid climate change – and there is general agreement by most people who have seriously looked at the problem that we should be embarking on mitigation efforts right now. Mitigation schemes will have little effect on the climate of the next few decades, but a successful mitigation effort would allow us to avoid large climate changes occurring in the second half of this century and beyond.

As we learned in Chapter 11, reducing greenhouse-gas emissions requires improvements in the energy efficiency of our economy and, most importantly, converting our energy system to one that primarily utilizes carbon-free energy sources, such as solar, wind, nuclear, and CCS. We also found out in that chapter why most economists agree that emissions reductions will not occur without government intervention. In this chapter, we explore in detail the policy options that governments can use to reduce emissions.

12.1 Conventional regulations

The conventional approach to regulation, often described colloquially as *command-and-control* regulation, requires all emitters in a particular economic sector to meet a single standard. Electricity companies, for example, might be required to generate energy by using a particular technology, such as wind or CCS. Cars might be required to be hybrids. Alternatively, regulations may limit total emissions of a pollutant, or enforce a standard of greenhouse gases emitted per kilowatt-hour generated (for power plants) or greenhouse gases emitted per mile driven (for cars).

The conventional approach has the advantage that it is clear and easy to understand. Even today, many environmental regulations, including regulations on air pollution, still fall into this category.

Since the 1980s, however, weaknesses with this approach have been identified and it has been falling out of favor with regulators. First, technologies specified (e.g., wind, CCS) may not actually turn out to be the best ones. Second, the regulations force all emitters to meet the same emissions standards. This ignores the fact that some emitters can reduce emissions more cheaply than others. Third, conventional regulations provide no incentive for the development of new technologies for emissions reductions beyond the specified target.

Because of these disadvantages, there is little talk in policy circles of attempting to solve the climate change problem with conventional regulations. Instead, market-based regulations are now preferred, and I will spend the rest of the chapter discussing them.

12.2 Market-based regulations

In Chapter 11, I discussed why the free market is unlikely to solve climate change without intervention from the government: The basic reason is that it is free to emit greenhouse gases into the atmosphere. These emissions impose a cost, but it is on the rest of the world, not the emitter. Economists call this an externality. Because the costs of the emissions are not imposed on the emitter, there is no incentive for the emitter to make any effort to reduce emissions. The solution to this problem is to make emitters pay for emitting greenhouse gases. In so doing, we provide an economic incentive to reduce emissions.

Making emitters pay for their emissions is a market-based solution. It does not tell anyone how much they can emit, or what technology to use – it only requires them to pay for whatever emissions they do make. In the next two sections, I will discuss the two market-based regulatory approaches most frequently discussed in the climate change policy debate: carbon taxes and cap-and-trade systems. I will show how market-based mechanisms reduce emissions at a lower cost than do conventional regulations.

12.2.1 Carbon tax

The first approach is a *carbon tax*. Under this policy, emitters have complete freedom to emit as many tons of greenhouse gas to the atmosphere as they choose, as long as they pay a specified fee to the government for each ton released to the atmosphere.

To understand why a carbon tax reduces emissions, let's imagine a power plant, which we will call Plant A, that emits 10 tons of carbon dioxide into the atmosphere each year. The third column of Table 12.1 contains the *marginal cost* for reducing emissions for Plant A, which is the cost of reducing a particular ton of emissions. Thus, Plant A can reduce its annual emissions by 1 ton – from 10 tons to 9 tons – for $1. Reducing emissions another ton, from 9 tons to 8 tons, costs an *additional* $2. Thus, the total cost of reducing emissions from 10 to 8 tons is the sum of the marginal costs,[1] that is, $1 + $2, for a total cost of $3. The next ton of emissions costs $3 to

---

[1] For those of you who know calculus, you can think of the marginal cost as the derivative of the total cost function and the total cost as the integral of the marginal cost function.

Table 12.1 Cost of reducing emissions for Plants A and B					
Emissions reduced by (tons)	Units emitted (tons)	Plant A's cost ($)		Plant B's cost ($)	
		Marginal	Total	Marginal	Total
0	10	–	–	–	–
1	9	1	1	2	2
2	8	2	3	4	6
3	7	3	6	6	12
4	6	4	10	8	20
5	5	5	15	10	30
6	4	6	21	12	42
7	3	7	28	14	56

eliminate, so the total cost of reducing emissions from 10 to 7 tons is $1 + $2 + $3 = $6. And so on.

The marginal cost of reducing emissions increases as emissions are progressively reduced. To see why, think about golf. When you just start out, you may be shooting 130 strokes over 18 holes. It takes relatively little effort to reduce your score by one stroke, down to 129 – perhaps only 1 hour at the driving range. Taking another stroke off your score, reducing your score to 128, takes slightly more work, maybe 2 more hours at the driving range. By the time your score reaches 80 and you're getting close to par, it can be extremely difficult and take enormous practice to reduce your score by one stroke to 79. Eventually, as you approach and possibly surpass par, you reach a limit beyond which you'll never move past, no matter how hard you work.

In this golf example, the marginal cost is the amount of effort you have to apply to reduce your golf score by one stroke. The fact that the amount of work it takes to drop your score by one stroke increases as your score diminishes is an example of what is often called the law of diminishing returns, which says that each additional unit (in this case, an improvement of one stroke in your golf score) costs more than the previous unit.

For Plant A, simply adjusting machinery in the plants may reduce the first ton at very little cost. Once the equipment has been tuned up, however, additional reductions are harder to make. Eliminating the second ton may require replacing some outdated equipment with newer, more efficient equipment. This will cost more than the first ton. The third ton may require even more equipment replacement, or perhaps wholesale changes in the plant's processes. This ton will therefore be more expensive to eliminate than the previous two tons.

Now imagine that a carbon tax of $4 per ton is imposed on the emitters – meaning that for every ton that is emitted into the atmosphere, the plants have to pay the government $4. How would each plant respond? Remember that Plant A has total freedom to emit as much as it wants – the carbon tax does not specify any reduction. Plant A will therefore search for the cheapest alternative.

Plant A can emit the 10th ton and pay a tax of $4, or it can pay $1 and not emit that ton. It does not take a financial genius to conclude that the rational thing to do is to not emit the ton. Now emissions are down to 9 tons, and Plant A can emit the

9th ton and pay a tax of $4, or it can pay $2 and not emit that ton. For this ton, too, the rational thing to do is to not emit that ton. Now emissions are down to 8 tons, and Plant A can emit the 8th ton and pay a tax of $4, or it can pay $3 and not emit that ton. Again, the rational thing to do is to not emit that ton.

Now things get a bit trickier. For the 7th ton, Plant A can emit the ton and pay a tax of $4 or it can pay $4 and not emit that ton. From a purely financial point of view, these two alternatives make no difference. I suspect, though, that most companies will reduce that last unit, because it's free to do so and the company can burnish its environmental reputation. So we can assume that Plant A will choose to not emit the 7th ton.

For the 6th ton, Plant A can emit the ton and pay a tax of $4 or it can pay $5 and not emit that ton. The rational thing to do in this situation is to pay the tax and emit that ton. Thus, under a carbon tax of $4, Plant A will reduce emissions by 4 tons.

Now let's consider a second plant, which we will call Plant B, and whose marginal costs are also listed in Table 12.1. Plant B can emit the 10th ton and pay a tax of $4, or it can pay $2 and not emit that ton. Clearly, Plant B will not emit the ton. Now emissions are down to 9 tons, and Plant B can emit the 9th ton and pay a tax of $4, or it can pay $4 and not emit that ton. As for Plant A, we can assume that Plant B will choose to not emit that ton in this case. Now emissions are down to 8 tons, and Plant B can emit the 8th ton and pay a tax of $4, or it can pay $6 and not emit that ton. Here, the rational thing to do is to pay the tax and emit that ton. Thus, under a carbon tax of $4, Plant B will reduce emissions by 2 tons.

Thus, Plant B reduces emissions less than Plant A does for the same carbon tax rate. The reason is that Plant B has higher marginal costs for reducing emissions. This may arise for any number of reasons – for example, Plant B may be older than Plant A and so be using outdated technology that is not amenable to reducing emissions.

To summarize, under a carbon tax each emitter will reduce emissions until the marginal cost of reduction is equal to the carbon tax rate. For Plant A, the marginal cost equals the tax rate of $4 per ton when emissions have been reduced 4 tons, whereas for Plant B, the marginal cost equals the tax rate when emissions have been reduced 2 tons. Because marginal costs vary among emitters, some emitters will make deeper cuts than others.

The total reduction in emissions from the two plants in response to a carbon tax of $4 per ton is 6 tons: a reduction of 4 tons from Plant A and 2 tons from Plant B. The total cost to Plant A of reducing emissions is $1 + $2 + $3 + $4 = $10, whereas the total cost to Plant B is $2 + $4 = $6. Thus, the total cost of the reducing 6 tons of emissions is $10 + $6 = $16.

Under a conventional command-and-control approach, there is a single performance target that each plant is required to meet. So an emissions reduction of 6 tons might be achieved, for example, by having both Plant A and B reduce emissions by three tons. This would cost Plant A $1 + $2 + $3 = $6 to reduce 3 tons, whereas it would cost Plant B $2 + $4 + $6 = $12 to reduce 3 tons. The total cost of this 6-ton reduction is $18.

This is an important result. The carbon tax of $4 per ton resulted in a 6-ton reduction for $16, whereas the conventional command-and-control approach resulted in a 6-ton reduction for a cost of $18. The carbon tax is cheaper because of its *flexibility* – it shifts reductions to the lowest marginal cost emitters, in this case, Plant A. In this way, emissions reductions under a carbon tax are made where they are cheapest, which lowers overall cost to society.

A carbon tax would be reasonably easy to implement. Most greenhouse gases come from fossil fuels, and these are produced at a relatively small number of sites. A carbon tax could be applied to the fossil fuel when it is extracted from the ground, using the administrative infrastructure for existing taxes, such as excise taxes on coal and petroleum. The price of the tax would then follow the fuel through the market, where the end user would finally pay it. A tax credit would be generated if the carbon is used in such a way that it was not released into the atmosphere (such as production of plastic or capture of carbon in coal combustion followed by sequestion).

As part of a long-term policy, the carbon tax would start out relatively small and, over several decades, gradually increase until emissions reached the target level. Gases other than carbon dioxide, such as methane or nitrous oxide, would also be taxed, but at a rate that takes into account how effective each one is at warming the planet. For example, 1 ton of methane contributes approximately 20 times more warming than a ton of carbon dioxide, so the tax on methane should be proportionately higher than the tax on carbon dioxide.

The costs of reducing emissions would eventually be passed on to consumers. Thus, the net effect of a carbon tax is to raise the prices of goods and services by an amount determined by the amount of the greenhouse gases released. Goods and services that are produced with little or no emission of greenhouse gases will not experience price increases, whereas the costs of goods and services that require the emission of significant amounts of greenhouse gases may see large prices increases.

Many people automatically consider taxes to be bad, so they look at suggestions of a carbon tax with, to put it mildly, disdain. However, most economists argue that a well-designed tax serves a useful economic purpose. For activities that generate negative externalities (costs imposed on society, such as emitting greenhouse gases or smoking cigarettes), a free market prices these activities too low, leading to overconsumption of the associated good or service. Taxes on these activities will provide correction for this and reduce consumption, which produces a more socially beneficial outcome. Thus, an economist thinks of a carbon tax as fixing a problem in the free market. Unfortunately, given the automatic opposition that new taxes generate, even those that make economic sense, the present prospects for implementing a carbon tax in the United States and in many other countries are dim.

12.2.2 Cap and trade

An alternative way to put a price on greenhouse-gas emissions is a *cap-and-trade* system. Under cap and trade, the government issues a fixed number of permits each year, with each permit allowing the holder to emit a fixed amount (often 1 ton) of greenhouse gas to the atmosphere. Emitters must hold permits for the amount of greenhouse gas they emit to the atmosphere. Thus, the total number of permits issued sets a cap on total emissions. Emitters with extra permits can sell them to those needing additional permits (hence the *trade* part). The price of the permits is set by the market, not by the government.

For the emitters, the economics of a cap-and-trade system are similar to the economics of the carbon tax. If the marginal cost of reducing 1 ton of greenhouse gas emissions is less than the cost of the permit, then the emitters will pay to not emit that ton. This allows them to either avoid having to buy a permit, or if they already have a permit for this ton, they can sell it at a profit. If the marginal cost is more than the permit, then the emitters will acquire a permit and emit that ton. In the end, the emitters will reduce emissions until the marginal cost of reducing emissions equals the price of the permits.

So if permits cost $4 per ton, then Plant A will use permits to emit 6 tons, thereby reducing emissions by 4 tons. Plant B will use permits to emit 8 tons, thereby reducing emissions by 2 tons. This is the same result that was obtained for a carbon tax of $4 per ton.

Under most cap-and-trade systems, when a unit of greenhouse gas is emitted, a permit is retired. Therefore, the government must continually issue new permits to replace those that have been used. Over time, the number of permits issued each year will decrease following a prescribed schedule covering several decades until the target emissions level is eventually reached.

One of the most contentious issues in any cap-and-trade system is how the government issues those permits. One approach is for the government to auction the permits off. In that case, companies would buy the permits from the governments and then pass the cost of the permits on to their customers through higher prices for their products. This approach has the advantage that permits go to those emitters who value the permits the most – and are therefore willing to pay the most. These will be the highest marginal cost emitters, for whom emissions reductions are most expensive.

This approach would also create an enormous transfer of wealth from consumers to the government. Some of this wealth could be used to help those with low incomes, who would be disproportionately hurt by the rise in energy prices that the cap-and-trade system would cause. The government could also use the income for other beneficial activities, such as research and development of new energy technology or reduction of the deficit. One frequently made suggestion is to use this money to reduce other taxes, such as those on labor and capital, or rebate the money evenly to every citizen. It should be noted that a carbon tax also creates a similar wealth transfer, so this issue would have to be tackled with that policy, as well.

Emitters, like utilities that burn coal, oppose auctioning the permits because they would have to pay the cost of the permits, which would then be passed on to consumers in the form of higher prices. This will reduce demand for their product – which would cost them money. The alternative is for the government to give away permits to companies for free. Because permits are as good as money (because they can be sold), this is equivalent to the government giving emitters money. Unsurprisingly, emitters favor this approach.

When giving the permits away for free, the decision about how to allocate permits is not determined by the market, like they would in an auction, but by other issues, such as fairness and political connections. For example, the imposition of a cap-and-trade system will be potentially disruptive to industries that emit a lot of carbon to the atmosphere (e.g., coal companies). Giving these industries free permits essentially provides financial aid to help them adjust to a new world where emitting carbon to the atmosphere is no longer free. Moreover, because free permits are equal to money, politicians can distribute them to curry favor from particular constituents or to buy support for the policy from particular industries.

In the most recent bill considered by the U.S. Congress in 2009 and 2010, the majority of permits would be given away for free initially, with a slow transition to auctioning 100% of the permits over the next two decades.

12.2.3 Carbon tax versus cap and trade

A carbon tax and cap-and-trade system are quite similar in many ways. Both reduce emissions by putting a price on emissions. Both systems allow companies to emit as much as they want, as long as they pay the tax or possess a permit for each unit emitted. In doing so, both move emissions reductions to where they are cheapest, namely to the lowest marginal cost emitters. In both cases, the emitter reduces emissions until the marginal cost of reducing the next ton of emissions is equal to the cost of the tax or permit.

Putting a price on emissions means that both approaches raise the price of fossil fuels and the goods and services made from them in exact proportion to the amount of greenhouse gases emitted by their consumption. Although consumers may not like to see prices go up, economics tells us it is the most efficient way to reach a socially optimum level of emissions, and it does so through several mechanisms. First, higher prices will encourage consumers to reduce their consumption of greenhouse-gas-intensive goods and services. Second, putting a price on emissions will encourage producers and consumers to substitute greenhouse-gas-free technology for their present technology. This occurs because these policies do not raise the prices of greenhouse-gas free technologies (e.g., wind, nuclear, and the products made from then), so they will make these more competitive. Third, and most importantly, it will encourage research and development of new technologies that can replace today's greenhouse-gas-intensive technologies. Humans are amazingly clever, and putting a price on emissions will signal the market that innovation and breakthroughs in greenhouse-gas-free technology will pay off handsomely. With this market incentive, we can expect new technologies that allow reductions of emissions to occur at the smallest cost.

However, there are some important differences between these two approaches. Under a carbon tax, the policy makers set the tax rate, which in turn sets the cost to society of the emissions reductions. But it is not exactly known what the economy's marginal cost of reduction is, so this means there is uncertainty in exactly how much of an emissions reduction will occur given a particular tax rate. Under a cap-and-trade system, in contrast, the policy makers set the total number of permits issued, and therefore the total emissions from the economy. However, the uncertainty in the marginal costs means that it is not known how much it will cost to reach the specified level of emissions.

Here is a simple analogy. Imagine you go to a store to buy some soda. You're given $50 and instructions to buy as much soda as you can. In that case, you know the total cost you will incur upfront ($50), but you don't know how much soda this will purchase. This is analogous to a carbon tax. This creates the potential problem that the tax rate set by the government will not achieve the desired emissions reductions. However, the importance of this problem is frequently overstated. Because emissions reductions will take several decades to reach the desired level, the tax rate can be adjusted over time so that the desired emissions levels are eventually reached.

Now imagine that you go to the store and are given instructions to buy 12 cases of soda. In this case, you know exactly how much soda you'll get (12 cases), but you don't know the cost. In the worst-case scenario, you might significantly underestimate the cost of a case of soda and not take enough money – and thereby be unable to get all 12 cases. This is the situation with a cap and trade. The total number of permits issued by the government sets the limit on emissions, so that is completely defined by the policy. However, the cost is uncertain. This creates a potential problem for cap-and-trade systems: Policy makers will issue too few permits (in an effort to bring emissions down sharply), and the cost of complying will be so high that significant economic disruption occurs. If this happens, the program would lose political support and might be abandoned.

To address the problem of runaway costs in a cap-and-trade system, an *escape valve* can be implemented. Should the cost of permits rise above a predetermined threshold, the government would sell more permits at that predetermined price. This would loosen the cap and increase emissions, but it would reduce the cost to the economy. If, in contrast, the government issues too many permits (in an effort to keep costs down) and the price of permits drops below a predetermined floor, then the government can commit to buying up all permits being sold at that price, thus preventing the price of the permits from falling below that floor.

For several political reasons, cap and trade has generally been the preferred climate policy for the last two decades. The European Union has a cap-and-trade system operating today. In the United States, however, political and economic events during the first half of the Obama Administration have made cap and trade a toxic commodity, and there seems to be no chance that an economy-wide one will be implemented at the federal level. In fact, at present it appears unlikely that the United States will pass any comprehensive mitigation policy — in the next few years, at least.

12.2.4 Offsets

Imagine that you own a power plant that emits 100 tons of carbon dioxide into the atmosphere every year. Imagine that you also plant a forest, which absorbs 100 tons of carbon dioxide from the atmosphere each year as it grows. If a carbon tax is implemented, how much tax should you pay? Do you pay a tax on emissions (100 tons, from the power plant) or do you get credit for the carbon dioxide removed from the atmosphere by the forest? If you get full credit for the forest, then your net emissions are zero because the emissions from the plant are canceled by the uptake by the forest and you would owe no carbon tax.

Actions taken to remove carbon dioxide from the atmosphere are known as *offsets*, which can be thought of as "negative emissions." Whether credit is given for offsets in a carbon tax or cap-and-trade regime, and how much credit, is one of the most contentious issues in the debate over climate policy. In particular, if credit is given for offsets, and these can be sold, then they become extremely valuable because they can be used to offset emissions.

From a physics standpoint, there is no question that offsets should count as "negative emissions": There is no difference to the climate between not emitting 1 ton of carbon dioxide and emitting 1 ton while, at the same time, removing 1 ton via an offsetting mechanism. Offsets also make sense from an efficiency standpoint. In much the same way that a carbon tax or cap-and-trade system are efficient because they encourage the lowest marginal cost emitters to make the reductions, offsets provide further flexibility in exactly how emissions are reduced. It may be cheaper, for example, for a coal-fired power plant to offset emissions by planting trees than it would be for it to capture carbon. Because of this, offsets would be expected to lower the total cost of reaching any specified emissions target.

However, practical issues make offsets a much more difficult proposition. First, many offsets are difficult to verify. For example, measuring carbon uptake by a forest is an extremely complex – perhaps even impossible – problem. And what happens if the forest grows for several years, and then a forest fire burns it to the ground, releasing the sequestered carbon dioxide back to the atmosphere? How is this accounted for? Does the forest owner have to refund the payments he received for the offsets?

Then comes the question of *additionality*. Additionality means that the offsetting action would not have taken place without the additional value given to the offsetting action by the carbon emissions regime. To understand what I mean by this, consider the following example. You own a plot of land, so you go to a local power plant and offer to plant trees on it if they pay you. They do so, and in turn they use the carbon absorbed by the forest to offset some of their emissions, which reduces their carbon tax.

The problem arises because we don't know what would have happened without the payment from the power plant. Perhaps you would have planted trees anyway, in which case the payments from the power plant did not lead to any reduction in carbon in the atmosphere. In order for these offsets to actually reduce carbon in the atmosphere, we must be sure that the offsets are *in addition* to what would have happened anyway and would not have occurred without the value that the offsetting actions have for climate change avoidance. Otherwise the offsets achieve no environmental good. Mitigation programs that include offsets must therefore establish a mechanism to determine whether an offset satisfies additionality.

Finally, there is the question of the morality of offsets. Offsets allow emitters to continue releasing greenhouse gases into the atmosphere by paying for reductions elsewhere. This has been compared to the indulgences purchased from the Catholic Church by the wealthy that forgave sin – and allowed them to continue sinning without any fear of divine retribution. The moral argument states that emitting greenhouse gases is fundamentally immoral and one should not be allowed to offset an immoral action.

For these reasons, offsets turn out to be one of the most controversial aspects of any mitigation program. In fact, some of the biggest stumbling blocks in the negotiation of the Kyoto Protocol were the proposals to allow offsets from forests and agricultural lands to satisfy a major part (between one fourth and one half) of the total emissions reductions of each country. This proposal was pushed by the United States (a country with a lot of forest and farm lands), but it was steadfastly rejected by Denmark and Germany.

12.3 Information and voluntary methods

A final way to reduce emissions is simply to give people information. If people can be convinced that climate change is a serious problem, and then provided ways to address the problem, they may take some action to address it without any further prompting by the government.

Information can indeed affect purchasing decisions. For example, car dealerships are required by law to put mileage stickers on cars they are selling. Although not every car buyer is concerned with mileage, many are and this information helps them make the socially preferred decision of buying a high-mileage car.

In the case of climate change, an example of relevant information is a greenhouse-gas registry. The requirement to simply report emissions can provide strong incentives for companies to reduce their emissions. Companies whose emissions far exceed those of their competitors will be embarrassed, whereas those with low emission may be viewed as socially responsible and thereby favored in the marketplace. In both cases, a registry will give companies incentive to reduce their emissions.

However, this approach has limits. Although informational and voluntary approaches are quite useful and can be effective at encouraging people to make some changes to their behavior, these approaches generally do not compel people to make large or difficult changes. For example, imagine that your professor asks you to work a few extra problems (that will not be graded) before the exam. Given the expectation that working the problems may help you on the exam, and that it's not a large investment in time, you might choose to work the problems. Now imagine that your professor asks you to write a 25-page term paper. If the paper represents a large fraction of your grade, then most students will take the assignment seriously. However, if your professor informs you that she is not going to grade the term paper, then many students will not put much effort into the paper.

So the government can provide the general public with information about climate change and information about how to reduce emissions, and it may well cause some people to make some changes. However, the large changes necessary for us to stabilize the climate are too big to be motivated simply because we've been told we *ought* to make those changes. Thus, informational and voluntary approaches will likely form part of our response to climate change. They will not, however, form the fundamental basis for our approach to reduce emissions.

12.4 Chapter summary

- The central pillar of most mitigation policies is putting a price on emissions of greenhouse gases. There are two primary policies to do this: a carbon tax and a cap-and-trade system.
- Under a carbon tax, emitters must pay a tax for each unit of greenhouse gas emitted. Under a cap-and-trade plan, each emitter must hold government-issued permits equal to the amount of greenhouse gases emitted; extra permits can be traded.
- Over decades, the tax rate will rise or the number of permits issued will decrease following a predetermined schedule until the desired emissions level is reached.
- Under these policies, emitters reduce emissions until the marginal cost (the cost of reducing the next unit) is equal to the carbon tax or the price of the permit. These policies are cheaper than conventional regulations because they shift emissions reductions to where they can be made most cheaply.
- Offsets are processes that remove carbon from the atmosphere – they can be thought of as negative emissions. Whether these are allowed to offset real emissions is one of the most contentious parts of emissions-reduction policy debates. Offsets should satisfy additionality for them to count. This means that the offsetting activity would not have occurred without the additional value of the activity from its impact on emissions.
- Because of the long lifetime of carbon dioxide, as well as the time it takes for mitigation policies to reduce emissions, mitigation efforts we begin today will significantly affect the climate only in the second half of the 21st century.

Terms

Additionality
Cap and trade
Carbon tax
Command-and-control regulation
Escape valve
Flexibility
Marginal cost
Offsets

Additional reading

There has been a huge amount written about both carbon taxes and cap and trade. A quick Google search will turn up more books, magazine articles, and whitepapers than you can ever read. So dive in! What follows here are a few particular suggestions.

P. Krugman, "Building a Green Economy," *New York Times Magazine*, April 7, 2010. This is a clear and concise summary of the economics of climate change policy. It pulls together many of the concepts from Chapters 11 and 12 (download at http://www.nytimes.com/2010/04/11/magazine/11Economy-t.html).

Richard Conniff, "Blue Sky Thinking: The Political History of Cap and Trade," *Smithsonian Magazine*, August 2009 (download at http://www.smithsonianmag.com/science-nature/Presence-of-Mind-Blue-Sky-Thinking.html). This article explains how the concept of cap-and-trade first emerged from the Reagan White House as a way for the free market to solve environmental problems. Times have indeed changed!

Congressional Budget Office, *The Economics of Climate Change: A Primer* (Washington, DC: CBO, April 2003). This is a valuable, although technical and slightly out of date, summary of the economics of climate change and our policy options (download at http://www.cbo.gov/doc.cfm?index=4171&type=0).

Problems

1. a) Explain how a carbon tax works.
   b) Explain how a cap-and-trade system works.
   c) What is the fundamental difference between these two policies?
   d) Given a carbon tax of x dollars (or a permit price of $$x$), an emitter will reduce emissions until what criterion is satisfied?

2. Why are emissions reductions achieved by use of a carbon tax or cap-and-trade system cheaper than those achieved by use of conventional regulations?

3. a) What is an offset?
   b) What does *additionality* mean?

4. In a *New York Times* op-ed piece (December 6, 2009), climate scientist Jim Hansen makes the following argument: "Consider the perverse effect cap and trade has on altruistic actions. Say you decide to buy a small, high-efficiency car. That reduces your emissions, but not your country's. Instead it allows somebody else to buy a bigger S.U.V. – because the total emissions are set by the cap." He argues that this renders a cap-and-trade system ineffective. Why is this argument wrong?

5. Imagine a carbon tax is implemented. One day, you decide not to drive to the grocery store, and you apply for offset credit for the emissions that did not occur because this trip was not taken. Should you get paid for this? What would you have to prove in order to get paid?

6. For the following, assume that Plants A and B have the following marginal costs for reducing emissions:

Number of units reduced	Marginal costs for Plant A	Marginal costs for Plant B
1	3	1
2	5	2
3	7	3
4	9	5
5	11	9

a) The government tells both plants to reduce three units of output. How much does this "conventional" regulation cost each plant? What's the total cost?

b) The government implements a carbon tax of $5 per unit. How much does each plant reduce? What is the total cost?

c) Which approach is cheaper? Why is the cheaper approach cheaper?

7. The table below shows the marginal costs of the following two plants, each of which emits 10 units each year. They both have six permits, meaning that each would have to reduce 4 units.

Number of units reduced	Marginal costs for Plant A	Marginal costs for Plant B
1	1	3
2	2	6
3	3	9
4	4	12
5	5	15
6	6	18
7	7	21

a) How many permits will Plant B buy from Plant A?

b) In what price range will these permits exchange hands?

8. Why won't voluntary and informational approaches lead to deep reductions in emissions?

13

A brief history of climate science and politics

In Chapters 11 and 12, we explored our options for addressing climate change. We can adapt to the change, we can mitigate it by reducing the emissions of greenhouse gases, or we can geoengineer the climate. In Chapter 14, I will pull all these together so we can explore how we can choose among these options. Before we get to that discussion, though, I describe the context of the policy debate by providing a brief history of climate change science, policy, and politics.

13.1 The beginning of climate science

People have been speculating on the nature of the climate for millennia, but modern climate science began in earnest two centuries ago, in the early 19th century. In the 1820s, mathematician Joseph Fourier provided one of the first descriptions of the physics of what we now know as the greenhouse effect: A planet's atmosphere can trap heat and warm the surface of the planet beyond what it would be if it were a bare, airless rock (we covered this physics in Chapter 4). Several decades later, in 1859, physicist John Tyndall discovered that it was primarily water vapor and carbon dioxide in the atmosphere that provided the warmth, despite that fact that these two constituents make up just a small fraction of the atmosphere.

The first recognition of climate change occurred in the 1830s, when geologist Louis Agassiz and others identified glacial debris scattered across Europe. They correctly concluded that northern Europe must have previously been covered by ice. This was an unanticipated discovery; prior to that time, everyone had simply assumed that the climate they experienced was what it had always been and always would be. This discovery of widespread ice ages showed that climate had changed in the past, and it certainly suggested that it could change again in the future. This motivated much of the scientific study of climate over the next century.

By the end of the 19th century, our knowledge of the climate system was advancing rapidly. In 1896, scientist Svante Arrhenius, a Nobel Prize winner famous for his studies of chemical reactions, estimated that doubling the amount of carbon dioxide in the atmosphere would raise the Earth's temperature 5–6 °C. His value is remarkably close to modern estimates of climate sensitivity, 2.0–4.5 °C, despite the fact that he had only the most rudimentary data about the climate system.

Although Arrhenius' calculations were primarily focused on explaining the ice ages, he also realized that humans were adding carbon dioxide to the atmosphere from coal combustion. He estimated, however, that it would take thousands of years before humans would emit enough carbon dioxide to significantly warm the climate.

He did not appreciate that fossil fuel use was growing exponentially. As we learned in Chapter 10, the growth of exponentials is so fast that it is easy to underestimate the long-term change.

The work of Arrhenius really marks the beginning of the theory of human-induced global warming. Nonetheless, although the bare outlines of modern climate science were apparent at the beginning of the 20th century, many fundamental aspects of climate science were still not well understood. Whereas Arrhenius had suggested that the carbon dioxide emitted by humans would accumulate in the atmosphere, many scientists thought that most of the carbon dioxide emitted by humans would be absorbed by the oceans (as discussed in Chapter 5, some carbon dioxide is indeed quickly absorbed by the ocean, but much is not). Furthermore, some scientists suggested that water so dominated the absorption of infrared radiation by the atmosphere that adding carbon dioxide would have no effect.

In addition, there was much less concern for environmental issues at that time. Nature was viewed as dangerous – in fairy tales, children who wandered into the woods risked their very lives. Indeed, the twists of weather and climate were among nature's cruelest weapons. When a tough winter or a severe drought could kill you, the battle against nature was a battle for survival.

So if the elements of human progress, such as the burning of fossil fuels, changed the climate, then that was okay. Cutting down forest and replacing it with farmland or hunting predators such as wolves to extinction were considered improvements. Nature was the enemy. Today, of course, we think differently about nature. We recognize that humans are strong enough to radically change nature, and we therefore view the wilderness as something to be protected and conserved (although sometimes we don't act that way). "Nature" is somewhere you may go on vacation, if you can find it and afford to travel there. Figure 13.1 schematically illustrates the power shift between nature and humans over the past century.

As an example, consider the fairy tale of Little Red Riding Hood. The end of that story features a woodsman's killing the wolf with an ax and then cutting the wolf open and rescuing the grandmother. Unlike today, no one in the 19th century felt bad for the wolf. A modern version of that fairy tale would end quite differently: Everyone would realize that it wasn't the wolf's fault it ate grandma, so biologists from the U.S. Fish and Wildlife Service would dart and tranquilize the wolf, extract grandma without injuring the wolf, then radio tag the wolf and release him into a national park – where he would live happily ever after and refrain from further consumption of grandmothers.

Temperatures rose during the first few decades of the 20th century (see Figure 2.2a), and by the 1930s it was apparent that the planet was warming. As *Time* magazine put it in 1939, "gaffers who claim that winters were harder when they were boys are quite right . . . weather men have no doubt that the world at least for the time being is growing warmer."[1]

Around this time, Guy Stewart Callendar, an English inventor and engineer, suggested that this warming was actually caused by human emissions of carbon dioxide. In his work he built off Arrhenius' observation that the burning of fossil fuels would

---

[1] Quoted in Weart (2003).

Fig. 13.1   Author's artistic impression of how people viewed their relationship with nature in the 19th century and today.

warm the planet, but he revisited old measurements of atmospheric carbon dioxide and, unlike Arrhenius, realized that humans were already increasing the global atmospheric level of carbon dioxide. Like most other people of this time, however, Callendar was not terribly worried about any detrimental effects of human modification of the environment.

13.2 The emergence of environmentalism

By the 1950s, our view of the environment was changing as a result of several factors. One was the invention of nuclear weapons. Nuclear bombs with a yield of several tens of kilotons had been used twice in World War II. By the 1950s, hydrogen-fusion bombs with yields hundreds of times larger had been developed. In fact, a single 1950s-era nuclear-armed bomber could carry bombs with more explosive power than all of the explosives used in World War II. It dawned on people that we humans now possessed the power to annihilate ourselves – and the idea that we could change the environment became more reasonable in comparison.

Air pollution was also becoming an important issue. Probably the most famous air-pollution event in history was the *killer smog* in London in 1952. In London at that time, most homes were heated with coal. In early December of 1952, a temperature inversion over London created a stagnant air mass over the city. As people burned coal, dark soot filled the air. This dark cloud hung over London, cutting out sunlight, which caused the temperature to plummet. This caused people to burn more coal for heat, leading to even more soot in the air. During the height of the event on Sunday, December 7, the visibility in London was 1 foot. Cattle in the city's market were killed and their carcasses discarded rather than sold because their lungs were black. The particulates harmed people's health and killed many of the weak and old. On December 9, the weather changed and the killer fog was blown away, vanishing as quickly as it had arrived – but not before several thousand Londoners died.

Air pollution in the United States was bad, too. My father was an undergraduate at Caltech, and he told me that visibility was so bad in Los Angeles when he arrived in the fall of 1948 that he was initially unaware that there were mountains nearby. After he'd been there a few weeks, a rainstorm blew through and enormous mountains near the campus were suddenly visible. He was, to put it mildly, surprised to see them. Such persistent air-pollution problems occurred throughout the world and convinced most people that the human impacts on the environment were not always for the better.

At the same time, people in many parts of the world were getting richer. People who are poor tend to worry about where their next meal is coming from or where they're going to sleep tonight – they are not terribly concerned with the environment. However, as people become richer and have disposable income to spend on less essential things, protecting the environment becomes a higher priority. Once you have money, you care about the view of the mountains, where you're going to go camping this weekend, and the extinction of polar bears.

Another important event was the International Geophysical Year, which took place in 1957 and 1958. This was an international effort that coordinated pole-to-pole observations of the Earth in order to improve our understanding of the fundamental geophysical processes that govern the environment. This intensive year of observations greatly improved our understanding of the Earth – and of the myriad of ways that humans can alter it. One of the most famous measurements started during the International Geophysical Year was of atmospheric carbon dioxide. Within just a few years after commencement, these measurements showed that atmospheric carbon dioxide was rising as a result of human activities (these measurements were plotted in Figure 5.6). Here was direct evidence of man's massive footprint on the planet.

Around this same time, large longitudinal studies were proving that smoking cigarettes was associated with various health risks, such as an increased risk of lung cancer. Given the possibility that these scientific results could hurt the companies' profitability, now-public documents from tobacco companies show that these companies responded by attacking the science. The goal was not to prove that cigarettes were safe but rather to create doubt, as described in a tobacco company document from 1969:

> Doubt is our product since it is the best means of competing with the "body of fact" that exists in the mind of the general public. It is also the means of establishing a controversy.[2]

One of the key parts of the doubt strategy was finding sympathetic scientists who would convey the message of doubt to the general public. Following this strategy,

---

[2] A large number of tobacco company documents can be viewed on the Legacy Tobacco Documents Library (see http://legacy.library.ucsf.edu/). This particular document can be found at http://legacy.library.ucsf.edu/tid/wjh13f00/pdf.

tobacco companies were able to successfully keep people confused and unsure about the exact state of the science for decades after the science had confidently established the connection. The success of the tobacco strategy encouraged others to adopt the strategy in other policy debates, including the debate over climate change.

13.3 The 1970s and 1980s: supersonic airliners, acid rain, and ozone depletion

Environmentalism may have begun in the 1950s, but several events in the 1970s solidified it in the general public's consciousness. Up first was the battle over the development of a supersonic airliner. Given the increase in speed of aircraft from the original Wright Flyer to the airliners such as the Boeing 707, it was generally believed in the 1960s and 1970s that the next step for commercial aviation was a supersonic airliner. Moreover, the country that developed the first successful supersonic airliner would garner enormous national pride as well as economic benefits. Because of this, Europe, the United States, and the Soviet Union engaged in a race to develop such an airliner in a competition much like the race to the moon.

By the 1970s, however, scientists realized that a fleet of supersonic airliners might have serious environmental consequences. Jet engine exhaust includes chemicals that can destroy ozone, and because supersonic airliners fly at high altitudes for efficiency, these effluents would be dumped directly into the stratospheric ozone layer. Scientists began to worry that, given a large enough fleet of these planes, a significant loss of ozone might result. Because stratospheric ozone blocks high-energy ultraviolet photons, which harm plants and causes skin cancer in humans, the loss of this ozone could have serious detrimental effects on the biosphere.

In the end, the Europeans developed their supersonic airliner, the Concorde, while the Soviet Union developed their version, the Tu-144. However, fewer than two dozen Concordes and Tu-144s were ever built. The Concordes flew for three decades, while the Tu-144s flew only a handful of commercial flights before they were removed from service. The United States completely abandoned its efforts to build a supersonic airliner. Although the concern over ozone depletion did play role in limiting the success of these supersonic planes, it was mainly economics; supersonic planes are expensive, and consumers would rather have cheap tickets than fast planes. In fact, today's modern planes, such as the Boeing 777, actually fly slower than a Boeing 707 – but they trade off speed for lower operating costs, allowing airlines to lower prices. As a result, fares in 2005 for flights from New York to Los Angeles – roughly $300 round trip – were about the same in nominal dollars as they were in the 1940s.

Just as the supersonic transport debate was subsiding, in 1973 and 1974, scientists first theorized that a class of industrial chemicals known as *chlorofluorocarbons*, known by their initials, CFCs, might deplete ozone. The issue of ozone depletion had already entered the public sphere during the supersonic airliner debate, so policy makers and the general public already had familiarity with this risk. At this time, the threat was completely theoretical – it would be a decade before actual observations of ozone depletion were obtained – and there was no effective replacement for CFCs in many applications. There were, however, some nonessential uses of CFCs. In the late 1970s, the United States banned CFCs from being used in one of these nonessential applications – as a propellant in aerosol spray cans.

In response to this, CFC manufacturers and industries that used CFCs in their products joined together to defend the product. The strategy they adopted would have been familiar to any tobacco executive: Hire scientists to make public statements casting doubt on the science. In doing so, they crafted a version of the tobacco strategy for use in environmental issues: It's not happening; if it is happening, we are not to blame; and, if we are to blame, then fixing the problem will be too expensive.

At this same time, another environmental problem was emerging: acid rain. Many power plants, particularly those that burn coal, emit large amounts of nitrogen oxides and sulfur dioxide to the atmosphere. Once in the atmosphere, these molecules can be absorbed by cloud droplets and raindrops, and once in liquid they react with water to form nitric acid and sulfuric acid. This is analogous to the way carbon dioxide dissolves into water to form carbonic acid (discussed in Chapter 5). The difference is that carbonic acid is what's known as a weak acid, whereas nitric and sulfuric acids are strong acids. This means that dissolving nitrogen or sulfur into water makes the resulting liquid much more acidic than dissolving an equal amount of carbon dioxide. When this acid rain falls to the ground, the types of potential damage it can do are numerous: bleaching of nutrients from soils, acidification of lakes and rivers, damage to wildlife and plants, damage to human-built structures, and so on.

This entire theory of acid rain is scientifically quite simple, and research done over the 1970s and 1980s definitively connected emissions from power plants to the acidic precipitation. In response to this research, the first broad international agreement covering acid rain, The Convention on Long Range Transboundary Air Pollution, was signed in Geneva by 34 member countries of the U.N. Economic Commission for Europe on November 16, 1979. The next year, the Council of the European Communities enacted a directive reducing sulfur dioxide emissions.

The Reagan Administration, however, was resistant to enacting regulations on emissions of acid-rain precursors. To support this decision, the Administration created a series of reports that argued that there was too much uncertainty to take action:

> "The state of the science, which allows healthy but abundant contradiction of scientific hypotheses, probably will not yield a scientifically complete assessment of acid deposition in the next few years," says the report prepared for Congress and the Reagan Administration. "To date, the state of the science will not allow assertive recommendations. Trends are weak and evasive. Data are spotty. . . . One of the most basic uncertainties is the extent of damage caused by acid deposition and its rate of change." The report says air pollutants, some of which cause acid rain, are the prime suspects in investigation of dying forests in the northeastern United States but "the association between damage and the occurrence of those pollutants is not well defined." It says some lakes have been acidified and fish have died but "the rate, character and full extent of these changes remain major scientific unknowns." And it says acidity can accelerate corrosion of buildings but "information about effects on materials from acid deposition needs to be better defined."[3]

---

[3] "Acid Rain Facts Called Sketchy," *The Globe and Mail* (Canada), June 12, 1984.

The United States took no action on acid rain during the Reagan Administration. However, in the early 1990s, the George H. W. Bush Administration enacted regulations to reduce sulfur emissions through a cap-and-trade system – just like the cap-and-trade systems discussed in Chapter 12. These regulations greatly reduced the emissions of sulfur at a price far below expectation. And, as expected, this greatly decreased the occurrence of acid rain. The success of cap and trade in helping solve the acid rain problem is one of the main reasons that policy makers looked hopefully at that mechanism for addressing climate change.

The ozone problem remained an active scientific and political issue throughout the early and mid-1980s. Scientists obtained the first evidence that the ozone was actually being depleted as a result of CFCs when they observed that, every spring, roughly 90% of the ozone over Antarctica was destroyed (it built back up during the rest of the year so that it was available to be destroyed again the following year). This was a shocking development, and it became known as the *ozone hole*. The original theories from the 1970s suggested that ozone depletion would be a slow process, taking half a century or longer for significant depletion of ozone to occur, and that it would primarily occur at mid-latitudes and high altitudes. The ozone depletion over Antarctica, however, was occurring much more rapidly than this, and in a different place than predicted.

Within a few years, newly discovered chemical reactions that relied on chlorine from CFCs combined with the unique meteorology of the polar regions were identified as the cause of this rapid polar ozone depletion. The observation of the Antarctic ozone hole confirmed the role of humans and suggested that this problem might be more dire than had previously been recognized. In response to this recognition, the world adopted the *Montreal Protocol* in 1988. This agreement committed the world to the phasing out CFCs over the following few decades.

An important aspect of the Montreal Protocol was that the phaseout of CFCs happened in two stages. Industrialized countries phased out CFCs first, followed 10 years later by developing countries. There are several reasons for this. Industrialized countries are richer than the developing countries, so they have more resources to apply to phasing out CFCs. Moreover, by having the rich countries go first, economies of scale and technical advances would be expected to drive down the cost for developing countries of phasing out CFCs. There were also ethical considerations. The CFCs in the atmosphere – which were causing the ozone depletion – had mainly been released to the atmosphere by activities in the industrialized countries. Developing countries had contributed little to the problem. All of this suggested that the industrialized countries had a responsibility to take the first step to clean up the problem.

Even as the science became more certain, so-called ozone skeptics stepped up their attacks on the science of ozone depletion. A good example is this quote from *National Review*:

> The current situation can fairly be summarized as follows: The CFC-ozone theory is quite incomplete and cannot as yet be relied on to make predictions. The natural sources of stratospheric ozone have not yet been delineated, theoretically or experimentally. The Antarctic ozone hole is ephemeral; it comes and goes, and seems to be controlled by climatic factors outside of human control rather than by CFCs.

The reported decline in global ozone may be an artifact of the analysis. Even if real, its cause may be related to the declining strength of solar activity rather than to CFCs. The steady increase in malignant melanoma has been going for at least 50 years and has nothing to do with ozone or CFCs. And the incidence of ordinary skin tumors has been greatly overstated. . . . And substituting for CFCs is no simple matter. A New York Times report of March 7, 1989 talks about the disadvantages of the CFC substitutes. They may be toxic, flammable, and corrosive; and they certainly won't work as well. They'll reduce the energy efficiency of appliances such as refrigerators, and they'll deteriorate, requiring frequent replacement. Nor is this all; about \$135 billion of equipment use CFCs in the U.S. alone, and much of this equipment will have to be replaced or modified to work well with the CFC substitutes. Eventually that will involve 100 million home refrigerators, the air-conditioners in 90 million cars, and the central air-conditioning plants in 100,000 large buildings. Good luck! The total costs haven't really been added up yet.[4]

In other words, according to this argument, 1) ozone depletion may not be happening; 2) if it is happening, it may not be CFCs; 3) if it is due to CFCs, then it's too expensive to do anything about. This argument is an argument about uncertainty, and it is a descendent of the strategy of the tobacco industry.

In retrospect, we now know that all of these arguments are wrong. Two decades of research have concretely verified the link between CFCs and stratospheric ozone depletion. What is more, the costs turned out to be so small that, when CFCs were completely phased out in the mid-1990s, no one noticed.

Why were scientists and politicians so worried about the CFC–ozone connection if the science supporting it was so shoddy? The same article provides an answer:

It's not difficult to understand some of the motivations. For scientists: recognition for keeping dusty records or running complicated computer models that are rather dull; more grants for research; press conferences; and newspaper stories. Also the feeling that maybe they are saving the world for future generations. For bureaucrats the rewards are obvious. For diplomats there are negotiations, initialing of agreements, and – the ultimate – ratification of treaties. It doesn't really much matter what the treaty is about, but it helps if it supports "good things." For all involved there is of course travel to pleasant places, good hotels, international fellowship, and more. It's certainly not a zero sum game.

I have left environmental activists to the last. There are well-intentioned individuals who are sincerely concerned about what they perceive as a critical danger to the health of future generations. Many of the professionals share the same incentives as government bureaucrats: status, salaries, perks and power. And then there are probably those with hidden agenda of their own – not just to "save the environment" but to change our economic system. The telltale signs are the attack on free enterprise, the corporation, the profit motive, the new technologies. Some are socialists, some are Luddites.

Most of these "compulsive utopians" have a great desire to regulate – on as large a scale as possible. To them global regulation is the "holy grail." That's what makes the CFC-ozone issue so attractive to them.

---

[4] S. F. Singer, "My Adventures in the Ozone Layer," *National Review*, June 1989.

The goal here is to create an alternative narrative for the issue: The science the public is hearing in the mainstream media is wrong, and the reason wrong science is being conveyed is that scientists and advocates are corrupt, biased, or stupid. No evidence is provided to support these charges, but it's not really necessary. The point here is to cast doubt, not to prove the accuracy of this narrative.

The Earth's temperature remained relatively constant between the 1940s and 1970s (Figure 2.2). Despite this, scientists in the 1970s remained focused on the potential problem of global warming caused by increasing abundances of atmospheric greenhouse gases. At the same time, however, the abundance of aerosols from human activities was also rising (e.g., the burning of coal containing sulfur). As discussed in Chapter 6, aerosols tend to cool the planet, which offsets some of the warming from increased greenhouse gases. Some scientists suggested that humans were in fact adding enough aerosols to the atmosphere to overpower greenhouse gases, and that the dominant influence of man was a net cooling of the climate.

A legitimate scientific debate ensued over which effect would dominate, and by the end of the 1970s the debate had been settled in favor of those predicting that global warming would be the dominant human influence. In recent years, some have misrepresented this debate to claim that the scientific community in the 1970s was predicting global cooling. This is incorrect – there was never any widespread consensus among scientists that aerosol-induced cooling was the dominant influence of humans.[5]

The 1970s ended with the publication of an influential report[6] from the U.S. National Academy of Sciences that reviewed the science and came to this conclusion: "If carbon dioxide continues to increase, the study group finds no reason to doubt that climate changes will result and no reason to believe that these changes will be negligible. The conclusions of prior studies have been generally reaffirmed." More quantitatively, they concluded that a doubling of carbon dioxide would result in a warming of 1.5–4.5 °C, which is very close to today's estimates. More research in the early 1980s fleshed out and confirmed the general view that humans were in the process of modifying the climate. However, the general public and most politicians were not focused on the issue.

13.4 The year everything changed: 1988

1988 was the year when climate change went from being a mostly academic problem to a political one. That summer was blisteringly hot, with much of the United States suffering under a drought and temperature records smashed on a seemingly daily basis. A small number of U.S. Congressional leaders were interested in the problem

---

[5] For a good review of the history of "global cooling" and how it is misrepresented in today's debate, see Peterson et al., "The Myth of the 1970s Global Cooling Scientific Consensus," *Bulletin of the American Meteorological Society* 89 (2008): 1325–1337.

[6] Ad Hoc Study Group on Carbon Dioxide and Climate, *Carbon Dioxide and Climate: A Scientific Assessment* (Washington, DC: Climate Research Board, National Research Council, 1979).

of climate change, and they felt the time was right to hold a Congressional hearing on it. In a stroke of political genius, they held the hearing in August, which is the hottest part of the Washington summer. Committee staffers opened the windows of the hearing room the night before, so the room itself was also extremely hot. Videos of the hearing show people sweating and mopping their brow, effectively reinforcing the message about climate change.

At that hearing, NASA climate scientist James Hansen declared that he was 99% confident that the world really was getting warmer and that there was a high degree of probability that it was due to human activities. Coming on the heels of the publicity over the ozone hole, this created a media firestorm, and it put the issue of climate change onto the political radar. In the next few months, the United Nations passed a resolution urging the "Protection of global climate for present and future generations of mankind." *Time* magazine, instead of naming a "Person of the Year" for 1988, named "Endangered Earth" the "Planet of the Year."

This was also the year that the Intergovernmental Panel on Climate Change, or IPCC, was formed (I discussed this organization in some detail at the end of Chapter 1). Its goal is to provide a summary of what we know about climate change and how confidently we know it for policy makers and the general public.

Along with this increased scientific awareness, people began thinking about reducing emissions of greenhouse gases. Given that the energy industry sold trillions of dollars of fossil-fuel-based energy every year, the prospect of regulations that would cost them some of these trillions motivated them to strenuously oppose regulations. They were joined by those philosophically opposed to environmental regulations. Vaclav Klaus, President of the Czech Republic and one of the very few leaders of any country to doubt the mainstream view of the science of climate change, sums their view up this way: "The largest threat to freedom, democracy, the market economy, and prosperity at the end of the 20th and at the beginning of the 21st century is no longer socialism. It is, instead, the ambitious, arrogant, unscrupulous ideology of environmentalism."[7]

With the success of the tobacco strategy in mind, those opposed to regulations on greenhouse gases chose the same tactic of attacking the science. Those leading the attack have become known as *climate skeptics*. Skeptics are a heterogeneous group, with some disputing all of these claims and others disputing only some parts; for example, some skeptics accept that the Earth is warming but dispute that humans are responsible. But they typically share a disdain for any science that might lead to increased government regulation. And some of the best-known climate skeptics are in fact the same people who also contested the science of acid rain and ozone depletion.[8]

In 1990, the IPCC put out its first assessment on the science of climate change. In it, the IPCC concluded that "the size of this [observed] warming is broadly consistent with predictions of climate models, but it is also of the same magnitude as natural climate variability. Thus the observed increase could be largely due to this natural variability." This relatively weak statement about the role of humans in climate

---

[7] V. Klaus, *Blue Planet in Green Shackles* (Washington, DC: Competitive Enterprise Institute, 2007).

[8] For example, Fred Singer, Sallie Baliunas, and Patrick Michaels.

change reflected legitimate uncertainties in climate science at that time. Because of these uncertainties, a definitive attribution of the warming to greenhouse gases, rather than solar variations and internal variations such as the El Niño/Southern Oscillation, was not possible.

13.5 The Framework Convention on Climate Change: The first climate treaty

Despite this, many world leaders felt that action had to be taken on climate change. The result was the Earth Summit in Rio de Janeiro in 1992. What emerged from that meeting was the treaty known as the *Framework Convention on Climate Change*, frequently referred to simply as the "framework convention" or by its initials, FCCC. The FCCC enjoys near-universal membership, with 192 countries having ratified it, including the United States, China, and all other big emitters. The principles enshrined in the FCCC remain the major building blocks on which negotiations of treaties to reduce emissions have been built.

The most contentious debate over climate change policies involves mitigation. In that regard, the stated goal of the FCCC is "to achieve stabilization of greenhouse gas concentrations in the atmosphere at a low enough level to prevent dangerous anthropogenic interference with the climate system." This is fine as far as it goes, and it receives widespread agreement at this level of abstraction. In practice, though, the meaning of this statement hinges on the definition of the word *dangerous*. There is no scientific definition of what dangerous climate change is because this is a value judgment – climate change that one person may perceive as dangerous may not be perceived that way by someone else.

In order to bring fairness or *equity* to any climate change agreement, the FCCC also enshrines the concept of *common but differentiated responsibilities*. This means that all countries must participate in solving the climate change problem, but not necessarily the same way. In particular, the rich, industrialized countries should begin cutting their emissions first, with developing nations cutting their emissions later.

The reasons for this are similar to the reasons that, in the Montreal Protocol, industrialized countries phased out CFCs first, followed 10 years later by developing countries. The industrialized countries of the world are far richer than the developing countries, so they have more resources to apply to reducing emissions. Moreover, by having the rich countries go first, economies of scale and technological advancement would bring down the cost of reducing emissions so that, when developing countries did begin reducing their emissions, the cost to them would be less.

There are also moral considerations. The 2 billion or so poorest people in the world currently live hard lives of crushing poverty. One of the ways to raise these people out of poverty is by economic growth – increasing their consumption of goods and services. This requires energy, so anything that makes consuming energy harder or more expensive for the poorest will also make it harder to lift these people out of poverty. Common but differentiated responsibility is a way of saying that solutions to climate change should not work at cross-purposes to efforts to reduce poverty.

There is also the question of historical responsibility. Most of the increase in carbon dioxide in the atmosphere over the past 250 years is due to emissions from the rich and industrialized countries. Because they are primarily responsible for the climate change that we have experienced, it makes sense for them to have a greater responsibility for taking the first steps toward addressing the problem. It is also clear, however, that developing countries must eventually contribute. China is now the largest emitter of carbon dioxide, and several other developing countries are either presently major emitters or on track to be. Reducing global emissions significantly would be impossible without these developing countries eventually making deep emissions reductions.

The FCCC also included what's known as the *precautionary principle*: "Where there are threats of serious or irreversible damage, lack of full scientific certainty should not be used as a reason for postponing such measures." This makes an important statement about the role of uncertainty in climate policy deliberations. Because the impacts of climate change are so potentially serious and irreversible, scientific uncertainty should not be used as an excuse to do nothing.

The FCCC was intended to be a starting point for more specific and binding measures to be negotiated later. Consequently, in contrast to its ambitious principles and objectives, the treaty's concrete measures were weak. Under the FCCC, nations committed to reporting their current and projected emissions and supporting climate research. Parties also accepted a general obligation to adopt, and report on, measures to limit emissions. What these measures had to be, or had to achieve, was not specified. Only for the industrialized countries did this general obligation also include the specific aim of returning emissions to 1990 levels by 2000. This target was nonbinding, meaning that it was an aspirational target and there were no sanctions for missing the target.

13.6 The Kyoto Protocol

By the middle of the 1990s, it was clear that no country would achieve the emissions-reduction target set in the FCCC and that a treaty with mandatory reductions would be required. Around that the same time, the IPCC released its second assessment of the science of climate change, in which it concluded that "the balance of evidence suggests a discernible human influence on the climate." This quote reflected the fact that the science of climate change had significantly advanced since the IPCC's first assessment and there was now much more evidence linking the observed warming to human activities. But significant uncertainties remained.

In response to these developments, the *Kyoto Protocol* was negotiated in December 1997. Unlike the FCCC's nonbinding emissions reductions, the Kyoto Protocol required emissions from participating industrialized countries, averaged over a commitment period running from 2008 through 2012, to be approximately 5% below their 1990 emissions level. Developing countries, on the other hand, did not have any emissions reduction requirements.

The Protocol incorporated several provisions to allow flexibility in how nations met their emission limits. As I discussed in Chapter 12, flexibility mechanisms allow emissions reductions to be shifted to where they can be made most cheaply, thereby reducing the overall cost of attaining a particular emissions target. Flexibility mechanisms included targets being defined for total emissions of a basket of carbon dioxide and five other greenhouse gases (methane, nitrous oxide, hydrofluorocarbons, perfluorocarbons, and sulfur hexafluoride). Countries could meet their target by reducing emissions of any of these gases, not just carbon dioxide.

Emissions-reduction obligations could also be exchanged between nations through various mechanisms. Under one mechanism, countries that can easily and cheaply reduce emissions can reduce their emissions below their target and sell their excess reductions to another country that is having difficulty reaching its target.

There are also project-based mechanisms. Under these mechanisms, an investor in one country may fund a project in another country that results in a reduction in emissions. For example, an investor in Europe could provide financing to build wind power in China (instead of a coal-fired power plant), and then get credit for the reduction in emissions associated with that project, which can be sold. Strict oversight of these programs is required to ensure additionality – for example, ensuring that the wind farm would not have been built without the European investor's funding.

The Protocol also included provisions for nations to meet some of their obligation through offsets. The Protocol included credit for reforestation, but some countries, such as the United States, wanted credit for other carbon-capturing activities, such as agriculture. This turned into a significant conflict at a negotiating session in November 2000 in The Hague. A proposed compromise was almost reached, but it was ultimately rejected at the last minute by the French and German environment ministers (both Green Party members), who judged that the proposed offsets weakened the Kyoto commitments too much.

13.7 The Bush years

Shortly after the breakdown in negotiations at the meeting in The Hague and just a few months after taking office in 2001, the Bush Administration announced it was withdrawing from the Kyoto Protocol process. The reasons included too much scientific uncertainty about climate change and potential harm to the U.S. economy. Although it later retreated from claiming that its withdrawal was based on scientific uncertainty, the Bush Administration continued to hold that the Protocol was unacceptable because of the high costs to the U.S. economy.

A particular problem cited by the Bush Administration was the absence of emission limits for developing countries. Although "common but differentiated responsibilities" is enshrined in the FCCC – which George H. W. Bush signed – and had also been incorporated into the Montreal Protocol, the George W. Bush Administration painted this as unfair to the United States. It is true that China was already a major economic competitor to the United States in many areas, but it is also true that China was still a much poorer country than the United States and had far fewer resources to devote to reducing emissions.

Also in 2001, the IPCC released its third assessment report on the science of climate change. The report came to this conclusion: "There is new and stronger evidence that most of the warming observed over the last 50 years is likely attributable to human activities." Compared with the previous reports, this one made a much more definitive statement about the role of humans in the recent warming. However, there was still uncertainty, as reflected by the use of the world *likely*. In the nuanced language of the IPCC, the word *likely* denotes a 2 out of 3 chance of being correct.

In February 2002, President Bush outlined his alternative approach to the issue. While consistently saying that climate change was a problem that needed to be addressed, the Bush Administration steadfastly avoided talking about reducing greenhouse-gas emissions. Instead, the emphasis was on reducing the greenhouse-gas intensity – the T term of the IPAT relation, which concerns emissions per dollar of GDP. Their stated goal was to reduce greenhouse-gas intensity by 18% by 2012. This was a weak goal because greenhouse-gas intensity has historically declined at 1–2% per year without any policies. Thus, the Bush goal would likely be met with little or no extra effort. In addition, given the growth in population and affluence, this reduction in greenhouse-gas intensity would have nonetheless accompanied an increase in total U.S. emissions.

The Bush policies also increased funding for climate change science and for specific technologies to reduce emissions, implemented tax incentives for renewable energy and high-efficiency vehicles, and started several programs to encourage voluntary emission cuts by businesses.

While the Bush Administration made no serious effort to reduce emissions, the rest of the industrialized world continued pushing for the Kyoto Protocol. The fate of the Protocol remained uncertain until late 2004. To enter into force – and so become binding on those who ratified – the Protocol required ratifications by 55 countries, including nations contributing at least 55% of 1990 industrialized-country emissions. This threshold meant that, without the United States, the treaty could enter into force only if all other major industrialized countries joined, including Russia. After several years of uncertainty about its intentions, Russia submitted its ratification in November 2004, allowing the Protocol to enter into force on February 16, 2005.

But the long delay awaiting the required ratifications meant that the Protocol entered into force only 3 years before the start of the commitment period. Nations' efforts in the meantime had been uneven, and many were not well prepared to achieve what would be a large deflection of emissions over very few years. As a result, national emissions trends have varied widely and few countries are likely to meet their commitments.

At the same time, in the absence of any U.S. federal efforts to reduce emissions, efforts in the United States trickled down to the state and local levels. For example, Connecticut, Delaware, Maine, Maryland, Massachusetts, New Hampshire, New Jersey, New York, Rhode Island, and Vermont banded together to form the Regional Greenhouse Gas Initiative, more commonly referred to by its initials, RGGI (and pronounced "reggie").[9] The RGGI is a regional cap-and-trade program that covers emissions from just one economic sector – electric power plants. The Western

---

[9] In May of 2011, New Jersey announced its intention to leave the program.

Climate Initiative is a similar group of western states and several Canadian provinces that are also developing a regional cap-and-trade system. In addition, many individual U.S. states and cities have also begun efforts to reduce emissions.

In 2007, the IPCC released its fourth assessment on the science of climate change and came to this conclusion: "Most of the observed increase in globally averaged temperatures since the mid-20th century is very likely due to the observed increase in anthropogenic greenhouse gas concentrations." This continued the trend toward stronger statements implicating humans in the warming – the words *very likely* here denote a 90% confidence that the statement is correct.

13.8 Copenhagen

With the end of the Kyoto Protocol in 2012 in sight, representatives of the world's governments met in Bali in December 2007 and agreed to negotiate a new climate treaty that would build on what the Kyoto Protocol had accomplished and that would be ready before a meeting in Copenhagen in 2009. Importantly, it was agreed in Bali that this new agreement would, unlike the Kyoto Protocol, include emissions reductions by both industrialized and developing countries. Subsequent negotiations quickly split over the relative efforts required of these two groups. By early 2009 it was clear that a new agreement would not be available by the time of the Copenhagen meeting.

As the December 2009 Copenhagen meeting approached, hopes were raised by renewed U.S. engagement under the Obama Administration. The Copenhagen meeting was marked, however, by continuing disputes over sharing the burden of action and many procedural roadblocks. Developing nations wanted the industrialized world to make sharp near-term (e.g., by 2020) reductions in emissions, whereas the industrialized world wanted the developing nations to agree to quantitative emissions reductions.

At the last minute, a "Copenhagen Accord" was negotiated that included several advances on any prior agreement. They agreed on a goal of limiting climate change to 2 °C (above pre-industrial temperatures), on emission cuts to achieve this, on verification of national actions, and on financial support for developing countries. However, the agreement was vague or weak on several key points. Moreover, objections from just five nations blocked the accord from formal adoption, so its status remains uncertain. It therefore remains unclear whether leaders of major nations are willing to act strongly enough to address the problem, or whether the current international negotiation process is able to motivate and coordinate such action.

As this book is written in mid-2011, a severe economic slowdown still grips the world economy. During recessions, people tend to adopt a shorter-term view of the world. They are more worried about paying the mortgage this month than they are about the climate in 100 years. As a result, there is little appetite in the U.S. or in China, the two biggest emitters, to enact legislation that would reduce emissions. Thus, what actions our society will take, and when, to mitigate the future effects of climate change remain in doubt.

13.9 Chapter summary

- Scientists have been studying climate change for nearly 200 years, and in that time a successful theory of climate has emerged. This theory is described in Chapters 1–7 of this book. Over the past 20 years, our understanding of the human impact on climate has greatly improved, and this is reflected in the evolution of the conclusions of subsequent IPCC reports.
- The first prediction of human-induced climate change was made by Svante Arrhenius, who recognized in the late 19th century that human combustion of fossil fuels might warm the climate. In the late 1930s, Guy Stewart Callendar made the first claim that human-induced global warming had arrived.
- In the 1950s, people realized that humans possessed the power to greatly modify our environment – and not to our benefit. And the economic growth and increases in wealth over that decade meant the environment had more value to people, and people had more money to spend to enjoy it.
- In the 1970s and 1980s, the debates over ozone depletion and acid rain were a preview for the debate over climate. Those opposed to action adopted the strategy of the tobacco companies: Cast doubt on the science. That explains why there is such vigorous disagreement over science in the public policy debate, even though there is widespread agreement among most climate scientists.
- The first climate treaty was the Framework Convention on Climate Change or FCCC. This treaty enshrined three important principles: 1) "common but differentiated responsibilities," 2) the precautionary principle, and 3) an agreement that the world should limit greenhouse-gas emissions in order to prevent "dangerous" climate change.
- In the mid-1990s, it became clear that emissions would not go down without a binding international agreement. The Kyoto Protocol included binding reductions of emissions for industrialized countries – these countries had to reduce emissions from 2008 through 2012 by roughly 5% below 1990 emissions. There were no restrictions placed on developing countries. The Kyoto Protocol entered into force in 2005.
- The question now facing the world is what happens after 2012, when the Kyoto Protocol ends.

Terms

Chlorofluorocarbons, or CFCs
Climate skeptics
Common but differentiated responsibilities
Equity
Framework Convention on Climate Change, or FCCC
Killer smog of London
Kyoto Protocol

Montreal Protocol
Ozone hole
Precautionary principle

Additional reading

S. R. Weart, *The Discovery of Global Warming*, 2nd ed. (Cambridge, MA: Harvard University Press, 2008). As I mentioned in Chapter 1, this is an accessible, well-written historical timeline of primary developments in the science of climate change, from the 1800s through the end of the 1900s with the formation of the modern consensus about the predominantly human cause of recent climate change as expressed in the 2001 IPCC report (accessible online at http://www.aip.org/history/climate/index.htm).

N. Oreskes and E. M. Conway, *Merchants of Doubt: How a Handful of Scientists Obscured the Truth on Issues from Tobacco Smoke to Global Warming* (London: Bloomsbury Press, 2010). As I stated in Chapter 1, this important book explains how deception is used to misguide the public on various matters, from the risks of smoking to ozone depletion to the reality of global warming.

R. Lizza, "As the World Burns: How the Senate and the White House Missed Their Best Chance to Deal with Climate Change, *The New Yorker*, October 11, 2010. This is a fascinating article about the ultimately unsuccessful attempt by three U.S. Senators to put together a climate change bill during the 2010 legislative session. It shows why it's so hard to get anything done on this problem.

Problems

1. Your roommates have a party when you are out of town.
   a) When you return, the apartment is a mess and they ask you to help clean it up. Do you help them?
   b) If you decide to help clean the apartment up, why? If not, what might they offer you to get you to help them?
   c) How is this situation analogous to the debate between developing and industrialized countries over mitigation efforts?
2. Who was the first person to discover that climate could change? Who was the first person to predict that human emissions of carbon dioxide might warm the climate? Who first claimed that human-induced climate change was occurring?
3. Do an Internet search and find some Web sites skeptical of mainstream climate science. List three of the claims they make about the science of climate change. Given what we've covered in the first 12 chapters of this book, are these arguments convincing?
4. What are the four important components of the Framework Convention on Climate Change?

5. What difference is there in how we view the environment today versus how people who lived in the 19th century viewed it? What are the factors that caused the change?

6. Explain the precautionary principle. Can you think of an example in your life where you've applied the concept (or explicitly not applied it)?

7. a) Explain the concept of equity as it was described in this chapter.

   b) How was it implemented in the Montreal Protocol?

   c) Give an example of how it might be implemented in a climate agreement.

14

Putting it together: A long-term policy to address climate change

We have now reached the final chapter on our trip through the problem of modern climate change. In the previous 13 chapters, we have explored the fundamental physics that leads us to confidently conclude that humans are now changing the climate, and that the continuing addition of greenhouse gases to the atmosphere will bring potentially significant changes to our climate over the next century and beyond. We are not certain how bad this climate change will be, but the upper end of the range (global and annual average warming of 5 °C or more by the end of the century) includes warming large enough for the experts to consider its impacts to be catastrophic. Even the lower end of the range (~2 °C) will be challenging for the world's poor as well as our most vulnerable ecosystems. We have also explored a number of possible responses to this risk, including mitigation, adaptation, and geoengineering.

The science of climate change plays an important role in helping refine decisions about what mix of these options we should pursue. However, science by itself does not determine policy. Policy decisions also involve value judgments concerning the relevant trade-offs among the options. Some of these trade-offs are economic and involve decisions about spending less money now or more money later, and some are moral, concerning the rights of future generations to inherit an unspoiled Earth versus the imperative to consume now in order to lift the poorest individuals out of poverty.

In this chapter, I will discuss the elements of an effective response to climate change. I will also show the step-by-step logic that underlies the most commonly suggested policies for addressing climate change.

14.1 Decisions under uncertainty: Should we reduce emissions?

The first thing we need to decide is whether we want to implement a mitigation program to reduce emissions, thereby avoiding some of the impacts of climate change that are expected to occur toward the end of the 21st century. This decision, however, must be made with incomplete knowledge because we do not know the answer to these key questions: How much warming will we experience if we do nothing? How bad will that much warming be? How expensive will it be to reduce emissions? How much warming can we avoid?

Despite the fact that we don't have precise answers to any of these questions, we must still make a decision now on whether to reduce emissions or not. Because of lags in the climate system and in our economy, we must begin efforts to reduce emissions now in order to significantly reduce warming in the second half of the 21st century. In any event, though, we have little capacity to affect the trajectory of temperatures over the next few decades.

This is not, I should emphasize, an unusual situation. Many important policy decisions must be made in the face of uncertainty. This includes important defense decisions (Should we invade Iraq?) and economic decisions (Should we cut taxes? Should we implement universal health care?), for example. Ultimately, decisions under uncertainty contain implicit value judgments about the choices. To better see this, consider the following two arguments:

> *Because the worst-case scenario of climate change is so serious, we must take action now to reduce emissions, even though we don't know exactly how bad climate change will be.*

> *Because of the high cost of reducing emissions, we must be certain that climate change is serious before we take action.*

Both statements argue that we must err on the side of caution in order to avoid a bad outcome. However, the bad outcome is different in these two arguments. In the first argument, the bad outcome is severe climate change, whereas in the second it is severe economic damage.

So which of these arguments is correct? We can gain some insight into how to think about this by looking at some familiar examples of decisions in the face of uncertainty. First, consider a criminal trial. To convict someone of a crime, a jury must be convinced that the defendant is guilty beyond a reasonable doubt. The reason for this standard is that we, as a society, have decided that it is better to acquit a guilty person than convict an innocent one.

Put another way, there are two errors a jury can make. They can convict an innocent person or they can acquit a guilty one. These errors are not equally bad – we judge that it is worse to convict an innocent person. So the standard of conviction ("guilty beyond a reasonable doubt") is set so that we minimize the possibility of making the worse mistake. In doing so, we increase the chance that we make the other mistake, acquitting a guilty person.

Another example occurs in deliberations concerning national defense. For example, former Vice President Dick Cheney famously said, "If there's a one-percent chance that Pakistani scientists are helping al Qaeda build or develop a nuclear weapon, we have to treat it as a certainty in terms of our response."[1] As in our jury example, there are two errors here that we could make. We could respond as if al Qaeda had a nuclear weapon, but it turns out they don't have a nuclear weapon. Or we don't respond, and it turns out they do have one. In most deliberations about national defense, it is generally felt that it is a worse error to be unprepared for a threat than it is to respond to a threat that never materializes. That is the fundamental judgment that Cheney is making here.

So how do we think about climate change? Must we be certain beyond a reasonable doubt that climate change is a serious threat to mankind before taking action to reduce

---

[1] Quoted in R. Susskind, *The One Percent Doctrine: Deep Inside America's Pursuit of Its Enemies since 9/11* (New York: Simon & Schuster, 2006).

emissions? Or should we take action to reduce emissions even if there's just a 1% chance that it is a serious threat? This question boils down to your judgment of which error is worse: reducing emissions unnecessarily because climate change turns out to be a minor threat or not reducing emissions and climate change turns out to be a serious threat.

Suppose that climate change turns out to be a minor problem. In that case, an aggressive mitigation program would impose costs as we pay to rebuild our energy infrastructure to switch from fossil fuels to renewables and other climate-safe energy sources. How bad would this be? Switching from fossil fuels to climate-safe energy has advantages completely unrelated to climate: energy security by reducing imports of oil from politically hostile countries, reductions in air pollution, and so on. More-over, because costs would be spread over the next several decades, at least some of the cost can be avoided by scaling back future efforts once we learn they are unnecessary. Furthermore, fossil fuels will be exhausted in the next century or so and thus switching away from fossil fuels is inevitable, so these costs are going to be paid eventually. The bottom line is that it is hard to imagine that a person living in Year 2100 is going to be very upset that we switched from fossil fuels to other energy sources.

Now suppose that climate change turns out to follow the worst-case scenario. If we do nothing to reduce emissions, then we doom the planet to much warmer temperatures for much of the next millennium. What is more, this warming might impose costs and risks that we would consider catastrophic – including the risk of abrupt or catastrophic shifts in climate. In this case, it is easy to imagine that a person living in Year 2100 will be furious that we did nothing to address a problem that we saw coming. In the end, it is difficult to make the argument that taking too much action on climate change is a worse error than taking too little. This would suggest a more symmetrical consideration of uncertainty — taking action based on the preponderance of evidence — or a standard closer to Cheney's, that the risk of climate change justifies action to reduce emissions, even in the face of significant uncertainty.

Another factor that enters into decisions under uncertainty is irreversibility. If an action you take is irreversible, then you have to be more certain that it's the right action than if a decision is easily reversible. That is why, for example, inmates on death row in the United States are allowed so many appeals to their death sentence – executing someone is as irreversible an action as there is, so you want to be as sure as possible that you're executing the right person. Just putting someone in jail, in contrast, is reversible – if you realize later that you've made a mistake, you can simply release that person.

Reducing emissions is a reversible decision. If we decide later that climate change is not that serious, then we can always change our policies and increase emissions of carbon dioxide. The investments we make in alternative energy sources, such as solar and wind, are reversible over a few decades.

But the converse is not true. If we continue emitting carbon dioxide, and then find out that climate change is a serious problem, there is no practical way to remove carbon dioxide from the atmosphere. Instead, we will have committed the planet to centuries of higher temperatures or centuries of geoengineering. And many of the impacts of climate change, like loss of the world's great ice sheets in Greenland and Antarctica, are essentially irreversible. Once we lose those ice sheets, they are not coming back until the Earth experiences another ice age. Other impacts, too, such as extinction of species, are absolutely irreversible. Thus, the irreversibility of emitted carbon dioxide and its associated climate change tends to favor taking action to reduce emissions. In the end, most people that have seriously looked at the problem, including just about every world government, have concluded that some mitigation effort is required.

14.2 Picking a long-term goal

If we decide that action should be taken to reduce emissions, we then have to decide how much to reduce emissions. The deeper the cuts in emissions, the less climate change we will eventually experience – but the more expensive those cuts will be. This is the trade-off, and we want to pick a target that avoids the worst climate change but at a cost that is manageable and does not interfere with other policy goals, such as poverty reduction.

The Framework Convention on Climate Change says that we should strive to avoid "dangerous" climate change, and at that level of abstraction the goal meets little resistance. But what is *dangerous*? It is not a scientific term, so science cannot tell us what is dangerous and what is not. Rather, it is a value judgment that takes into account our views on topics such as risk, poverty, environmental stewardship, government regulation, and many other contentious topics. The Kyoto Protocol specified only a short-term goal (approximately a 5% reduction below 1990 emissions of developed countries by the years 2008–2012). There are various ways to determine a long-term goal, and I discuss two of them in this section.

14.2.1 Cost versus benefits

In Chapter 13 we described how a company would respond if a price were put on emissions through a carbon tax or cap-and-trade system: It would reduce emissions until the cost of reducing one more unit (the marginal cost) equaled the cost of emitting that unit (either the tax on that unit or the cost of a permit to emit that unit). In this way, the company is pricing its alternatives and choosing the lowest cost option.

In an analogous way, we can quantitatively calculate how deeply to reduce emissions for the entire society by comparing the cost of various levels of emissions reduction to the benefits obtained by making those reductions. Imagine, for example, that our economy can reduce emissions of carbon by 1 ton for $5, but we get $50 of benefits by avoiding the climate impacts of that ton. From a societal standpoint, that's a no-brainer – we should certainly not emit that ton. Now imagine that the next ton of emissions can be avoided for $6, and avoiding emission of this ton delivers $49 of benefits. Again, it is clear that we should not emit that ton either. As we cut emissions deeper, the cost of eliminating each subsequent ton (the marginal cost) rises, while the benefits from avoiding that ton (the marginal benefit) decline. This is a manifestation of the law of diminishing returns, which we explored in Chapter 12.

The reduction that gives us the largest net benefit (benefits minus cost) would be our preferred goal. This occurs if we reduce emissions until the cost of reducing one more ton of emissions equals the benefit from avoiding that ton. Reducing emissions beyond that point is not cost effective because the cost of reducing that extra ton exceeds the benefit of not emitting it.

Even though this may be conceptually straightforward, the actual calculation is not. For example, we know what technology we have available right now to reduce emissions, and we can estimate the cost of reducing emissions based on today's best available technology. However, we also know that putting a price on carbon will spur development of new technologies by providing a financial incentive to reduce emissions that does not exist in today's economy. That this will happen is not in doubt – the question is how fast the innovation will take place. If we assume that innovation responds rapidly to putting a price on emissions, then the cost of reducing emissions will be much lower than if we don't make that assumption. Depending on what assumptions are made for this and other uncertainties, the resulting estimated costs of reducing emissions cover a wide range; some analyses conclude it will be quite cheap whereas others conclude it will be ruinously expensive.

Estimating the costs of the impacts of climate change – and therefore the benefits of avoiding it – is even more difficult. First, we do not at present have the ability to predict changes in temperature and precipitation at the regional scales required for detailed estimates of impacts. Second, converting estimated changes in climate into a dollar figure can be difficult and arbitrary. For goods and services that are traded in markets (e.g., food, lumber, recreational skiing), calculating economic loss that is due to climate change can be relatively straightforward. For things that are not traded in markets, however, estimates of the cost of climate change must be viewed with caution. Take, for example, the extinction of polar bears. Polar bears do not contribute much to the global economy, so their loss would have a negligible financial cost. But many people still value polar bears and would view their loss as a significant negative impact. Economic analyses can attempt to quantify the value of polar bears by using methodology that is neither uniform nor entirely satisfactory, or it can ignore their loss, which implicitly assigns it a value of zero – which is also not satisfactory. Given the many problems in estimating the cost of climate change, it should come as no surprise that there is also a wide range of estimates of its cost.

Another problem in estimating the costs of climate change comes from the timing of climate impacts. If 1 ton of carbon is emitted into the atmosphere today, it will warm the planet for centuries to come and will cause impacts over that entire time. However, the cost of not emitting that ton must be paid today. To compare costs that are occurring at different times, we therefore need to convert the value of the climate impacts over the next few hundred years to its value to us today (i.e., its present value). This involves discounting, which we explored in detail in Section 10.4.

Briefly, discounting accounts for the fact that money loses value as it recedes into the future, and we can use this concept to calculate the value to us today of costs and benefits occurring at different times in the future.

Although the discounting is conceptually straightforward, the big uncertainty is what discount rate to use. Most analyses in the climate change policy debate use discount rates between 0% and 4%. The larger the discount rate, the lower the present value of future climate impacts will be when discounted to today. We will consequently be wiling to pay less to avoid those impacts.

For example, with a discount rate of 0%, $1 trillion of climate change damage in 100 years has a present value of $1 trillion. We should therefore be willing to pay up to $1 trillion dollars today to avoid it. In contrast, if we select a discount rate of 4%, then $1 trillion of climate change damage in 100 years has a present value of $19.8 billion – meaning we would only be willing to pay $19.8 billion to avoid those impacts. Thus, the discount rate has an enormous impact on what we calculate our optimal climate policy should be.

This is particularly problematic for impacts occurring hundreds of years in the future. As we explored in Chapter 8, our emissions this century will commit the planet to warming over the coming millennium and beyond. So the present value of $1 quadrillion ($1,000 trillion) in 1,000 years at a 0% discount rate is $1 quadrillion. On the other hand, using a 4% discount rate yields a present value below 1 cent. Neither of these answers seems satisfactory, and there is no agreement among economists about exactly what the right discount rate to use is.

The uncertainties in estimates of the costs of reducing emissions, the costs of climate impacts, and the discount rate leads to a wide range of estimates of how deeply to cut emissions in order to maximize net benefits. Because of this, economics is limited in its ability to prescribe a quantitative response to climate change. Nonetheless, there are some things that all economic analyses agree on – and on those points we can have high confidence. There is widespread agreement that some reductions in emissions makes sense, and that that action should have as a centerpiece a price on emissions of carbon released to the atmosphere. In addition, there is agreement that the price on emissions should rise with time, eventually becoming very high. The net result is that emissions can be allowed to grow in the near term (for perhaps a decade or so), before emissions must actually begin to decline. The exact timing of the emissions peak is determined by how far we decide emissions must be cut – the more climate change we want to avoid, the nearer in time the cuts must begin. It's also important that we have near-universal participation in any climate regime, in order to make the necessary reductions at the lowest cost.

Economic analyses also suffer from another problem: It only looks at aggregate costs and benefits, but not their distribution. Imagine, for example, that an economic analysis finds that the present-day value of all climate impacts is $20 trillion, and that the optimal policy is to avoid $10 trillion of these impacts (leaving the other $10 trillion of impacts to occur, because avoiding them would cost more than $10 trillion). This type of analysis tells us nothing about distributional issues. Climate impacts will not be distributed evenly across the globe, and some regions will be hit harder than other regions. The best guess of climate science is that many of the hardest-hit regions are also the poorest regions of the world – regions that have contributed little to climate change and have the least resources available to address climate. Many would view allowing the poor of the world to bear the brunt of the unavoided climate change is fundamentally unfair. So although economics tells us what the most efficient solution is – in aggregate – it tells us nothing about whether the solution is fair or just to individual inhabitants of the Earth.

A final problem with economic analysis comes from the problem of catastrophe. Economic analyses tend to analyze the outcomes that are most likely. But sometimes unlikely things happen, and in the case of climate change the unlikely events might include some very, very bad outcomes. In particular, the worst-case scenarios for climate change over the next few centuries might include catastrophic outcomes such as abrupt climate changes, mass starvation, or even human extinction. Economic analyses are generally unable to assign a monetary value to a small but hard-to-quantify risk of truly terrible outcomes such as like these. For many people, uncertainty in how bad things can get trumps any quantitative cost–benefit analysis. For them, the mere possibility of a true catastrophe in the next century or so compels aggressive action to reduce climate change.

14.2.2 Target: 2 °C

Given all of the problems with economic analyses, expecting them to unambiguously determine the optimal amount of mitigation is not realistic. A simpler way to select a long-term goal is to simply pick a limit for temperature or atmospheric carbon dioxide above which you judge the climate impacts to be unacceptable. The limit should be low enough that it gives us a good chance to avoid serious climate impacts, but high enough that it is politically and economically acceptable. Over the past few years a consensus has grown up around a target of 2 °C of warming above pre-industrial temperatures. This is what was adopted, for example, at the FCCC's Copenhagen meeting in 2009. This is a challenging target – we have already experienced approximately 0.7 °C of warming above the pre-industrial level, and we are already committed to another roughly 0.5 °C of warming from emissions that have already occurred.

It is important to remember that this 2 °C is a political compromise. There is no scientific analysis that proves 2 °C is the most appropriate target, nor any reason to think that warming slightly below this threshold is much better than warming slightly above. In fact, it is possible to look at the fundamental trade-offs and come up with other judgments of where the limit should be. Some advocates, for example, argue that atmospheric carbon dioxide should not exceed 450 ppm. Others argue[2] that we've already passed the safe amount of carbon dioxide in the atmosphere and need to reduce it as quickly as possible to 350 ppm – roughly 40 ppm *below* the atmospheric abundance in late 2010. That is an incredibly challenging target, and it may require carbon-cycle engineering in order to accelerate removal of carbon dioxide from the atmosphere. However, as of the end of 2010, the target of 2 °C has most of the momentum in policy discussions.

---

[2] See, e.g., http://www.350.org/.

14.3 How do we get there?

14.3.1 The physics of a 2 °C limit

Let's assume that our goal is to limit warming to 2 °C above pre-industrial temperatures. In Chapter 6, we learned that our best estimate of the climate sensitivity is $0.75\,°C/(W/m^2)$, which means that for every watt per square meter of radiative forcing, the global average temperature increases by $0.75\,°C$. A limit of 2 °C temperature increase there means that we must limit radiative forcing that is due to human activities to $2.7\,W/m^2$. We also saw in Chapter 6 that humans have already applied a net radiative forcing of $+1.6\,W/m^2$ to the planet, meaning we are already committed to an eventual warming of roughly 1.2 °C (0.7 °C has already occurred and another 0.5 °C is committed).

The radiative forcing of $+1.6\,W/m^2$ arises from $+3\,W/m^2$ of radiative forcing from carbon dioxide and other greenhouse gases, a negative radiative forcing of $-1.2\,W/m^2$ from aerosols, plus a few other small factors. Thus, without the cooling effects of aerosols, we would have already exceeded the radiative forcing limit for keeping warming below 2 °C.

In order to keep radiative forcing as low as possible, we obviously need to rapidly decrease emissions of carbon dioxide. But given how much we rely on fossil fuels, and how much we have as a society invested in fossil fuels (e.g., gasoline-burning cars, coal-fired power plants), "as quickly as possible" turns out to not be all that fast. It seems likely that, even with an aggressive mitigation policy, it will take decades to significantly reduce emissions. Such a trajectory would see atmospheric carbon dioxide abundances continue to rise for several decades, eventually leveling off near 450 ppm. This would cause the radiative forcing from carbon dioxide to rise from $+1.6\,W/m^2$ today to $+2.2\,W/m^2$.

Aerosols are one of the primary components of photochemical smog and cause enormous health problems to those people who have to breathe them (e.g., the killer smog of London, discussed in Chapter 13). Because most people don't like air pollution, governments are trying to clean up the air – meaning reducing the emissions of aerosols and aerosol precursors. And because the lifetime of these aerosols is short (1 month or so), such efforts will lead to rapid reductions in their atmospheric abundance and rapid increases in radiative forcing. Our best estimates are that, because of the reduction of aerosols, radiative forcing will increase by $+0.8\,W/m^2$ over the next few decades.

Thus, increases in carbon dioxide and reductions in aerosols will cause radiative forcing to increase from $+1.6\,W/m^2$ to $+3.0\,W/m^2$. In order to keep net radiative forcing below $+2.7\,W/m^2$, we must therefore aggressively reduce emissions of other greenhouse agents – in particular, methane, ozone precursors, and black carbon. Ozone and black carbon are removed from the atmosphere in just 1 month or so, so reductions in the precursors that produce ozone or in emissions of black carbon will lead to almost immediate reductions in the atmospheric abundance of ozone and black carbon, and therefore radiative forcing. Methane has an atmospheric lifetime of a decade, so reductions in methane emissions will lead to reductions in radiative forcing in a decade or two.

An aggressive program of reductions in these short-lived greenhouse agents could lead to reductions in radiative forcing of as much as 1.3 W/m^2, although a more practical goal would be to reduce radiative forcing by approximately 1 W/m^2.

To summarize, here is the situation we're in: We want to limit radiative forcing to +2.7 W/m^2, which is approximately 1.1 W/m^2 above today's value of +1.6 W/m^2. Even with aggressive reductions of carbon dioxide emissions, the radiative forcing from carbon dioxide is still expected to increase by +0.6 W/m^2. In addition, strong action around the world to reduce air pollution will reduce the abundance of reflecting aerosols, which will further increase radiative forcing by +0.8 W/m^2. These changes will increase radiative forcing by +1.4 W/m^2, which would break our total radiative forcing limit of +2.7 W/m^2. Thus, we must also take action to reduce emissions of short-lived greenhouse agents (e.g., methane, ozone, and black carbon aerosols), which could give us enough of a reduction in radiative forcing to allow us to stay under our limit.

If we cannot get the radiative forcing low enough through mitigation efforts, then we might have to resort to geoengineering approaches, such as adding sulfate aerosols to the stratosphere, to temporarily reduce net radiative forcing and give mitigation efforts time to get the radiative forcing down far enough to keep temperatures from getting too high.

Note that we have assumed here a middle-of-the-road value for climate sensitivity of 0.75 °C/(W/m^2). As we learned in Chapter 6, there is uncertainty in this value, and the actual climate sensitivity may be bigger or smaller than this. In order to maintain our 2 °C limit on warming, bigger cuts would be required if we find out that the climate sensitivity is larger than this, whereas less stringent cuts would be required if we find that the climate sensitivity is smaller.

14.3.2 How to get there

As described in the previous section, we know what needs to happen from a scientific viewpoint in order to stabilize the climate at 2 °C above the pre-industrial level. The real question is fundamentally political: How do we get there from here?

Before we get to that, though, it's worth considering how big a challenge this is going to be. In order to cut emissions by 50–80% in 40 years, we would need to reduce emissions by, on average, 1–2% per year. In Chapter 8, we learned about the factors that control emissions: population, affluence, and technology. Over the next 40 years, most economists expect both the population and affluence of the world to increase. A rough estimate of population growth averaged over the next 50 years would be, say, 1% per year, while world affluence is projected to grow at 2–3% per year. In order for emissions to decrease at 1–2% per year while population and affluence are growing at their projected rates, we would need to reduce the greenhouse-gas intensity (the T term in the IPAT relation) by 4–6% per year.

This is an enormous challenge. Part of this will come from improvements in energy intensity (i.e., improvements in efficiency), which will probably contribute approximately 1% per year. Mainly, however, achieving high rates of reduction of greenhouse-gas intensity requires deep reductions in carbon intensity – that is, rapidly switching to energy sources that do not release greenhouse gases. Moreover, the amount of carbon-free energy required is immense. We would need to construct approximately 1 billion watts (1 GW) of carbon-free power every day between 2010 and 2050 to meet this target. To give you some idea of scale, a typical coal or nuclear power plant generates roughly 1 GW. This is not an impossible challenge, just a hard one.

Luckily, the broad outlines of what we need to do are clear. The first and most important action is to put a price on emissions of carbon dioxide and other greenhouse gases (why this is necessary was covered in Chapter 11). Economist William Nordhaus put it this way:

> Whether someone is serious about tackling the global warming problem can be readily gauged by listening to what he or she says about the carbon price. Suppose you hear a public figure who speaks eloquently of the perils of global warming and proposes that the nation should move urgently to slow climate change. Suppose that person proposes regulating the fuel efficiency of cars, or requiring high-efficiency light bulbs, or subsidizing ethanol, or providing research support for solar power – but nowhere does the proposal raise the price of carbon. You should conclude that the proposal is not really serious and does not recognize the central economic message about how to slow climate change. To a first approximation, raising the price of carbon is a necessary and sufficient step for tackling global warming. The rest is at best rhetoric and may actually be harmful in inducing economic inefficiencies.[3]

Second, although a price on carbon is crucial, there are some economic sectors where great progress can be made rapidly, but where expected progress with just a price on carbon will be slow. For such sectors, efficiency standards and other incentives can be implemented to encourage careful energy use. In addition, restrictions on activities that are particularly unfriendly to the climate, such as the burning of coal, might also be considered.

Third, fund the research and development of new technologies. Although it may be possible to solve the climate problem with existing technology, it is also clear that new technological developments can ease the transition as well as reduce the cost. Given this, experts agree that we, as a society, have not invested sufficiently in research into new technologies. As a result, investment in research should produce huge dividends and should be part of any policy to reduce emissions.

Fourth, prepare to adapt to climate change. Regardless of what actions we take now, the globe will continue to warm for decades. And to the extent that this warming cannot be avoided, then we must adapt to it. Thus, adaptation must necessarily be a part of our response to climate change. Anticipatory adaptation is cheapest, so we should begin to incorporate the reality of climate change into any plans for the future. This is particularly true for investments in long-lived infrastructure. When we build a road, airport, power plant, and the like, we must make sure that it is resilient to any reasonable changes in the climate. In addition, not every country has the resources to adapt, so mechanisms of providing international aid may have to be developed.

---

[3] See Nordhaus (2008), p. 22.

Fifth, there is also the unfortunate possibility that the world will not get its act together to reduce emissions. In such a situation, we will be faced with the choice of unrestrained climate change or geoengineering our way out of the problem. Because we may face this choice, we should prepare in advance by researching the geoengineering options in order to determine which one is most likely to work and what the possible negative side effects would be.

Sixth, realize that whatever policy we adopt now will probably not be exactly the right long-term policy. Because of this, policies must be reviewed and amended as new information about the science of climate change and new technological developments arise.

Unfortunately, although the requirements of stabilizing the climate are reasonably clear, the world has made essentially no progress toward achieving them. This is not because there is debate about whether stabilizing the climate is a reasonable goal, or because of scientific uncertainty – virtually every world government and world leader accepts the mainstream view of climate change espoused in this book, and they also accept that we know enough now to take action on climate change, even if we don't know everything.

The reason for inaction is fear over the costs.[4] So what are the short-term costs of reducing emissions? As already described, there are many estimates of this quantity from economists, and they cover a wide range from near zero to costs so ruinously high that they would move our standard of living back centuries. The highest and lowest estimates should be viewed with suspicion – some of these are purposefully designed to get a particular answer in order to advocate for or against efforts to reduce emissions. Even eliminating these, however, does not provide a high-confidence estimate of the costs.

In the end, we do not confidently know how much it will cost to reduce our emissions and make the transition to a fossil-free future. In this way, the climate change challenge is not unique; we almost never know in advance how much it will cost to comply with environmental regulations. Before regulations to reduce ozone depletion were passed in the 1980s, for example, some advocates were predicting that the regulations would cause an economic apocalypse, with people in the developed world having to get rid of their air conditioners and millions in the developing world dying because of a lack of food refrigeration.

It turned out that the cost of complying with ozone regulations was very inexpensive. Innovation caused by the threat of regulation led to the development of a substitute for the chemicals that destroy ozone that was so cheap that, when the new chemicals replaced the older ones, almost no one noticed. The hope with climate change is that, once a price is put on carbon emissions, energy efficiency will improve and innovation by the private sector will produce breakthroughs that allow us to make reductions in emissions cheaply and with minimal economic disruption.

Whether this will happen or not is impossible to know until we put a price on carbon. That's why I support[5] action to reduce emissions – even if the goal is modest.

---

[4] This is the primary concern at the government level. At the individual level, there are a host of concerns, from cost, to suspicion of the science, to the fear that environmental regulations will lead to socialism.

[5] I'm speaking here as a private citizen and not as a scientist.

If it turns out that reducing emissions is too much of a hardship economically, then the policies can be reversed and we can return to consuming fossil fuels without regard for the climate. But if reducing emissions turns out to have an acceptable cost, then we will be on the road to heading off a possible climate catastrophe.

14.3.3 What can you do?

If you have decided that climate change is something that we need to address, then you may be wondering what you as an individual can do. There certainly are personal choices you can make that will reduce your share of emissions of greenhouse gases. Some of these choices will not only reduce emissions, but also save you money and benefit you in other ways (e.g., get exercise by walking a mile to a friend's house instead of driving). Other choices may require upfront costs, but will pay for themselves over subsequent years (e.g., add insulation to your attic or switch to LED lighting). Some choices are difficult to justify on economic grounds alone (e.g., install photovoltaic solar panels on your house[6]).

But voluntary individual action is unlikely to motivate the emissions reductions necessary to stabilize the climate. Such reductions will require collective, coordinated action by society as a whole. That's why the single most important thing you can do is vote for politicians who support action on climate. Even better would be to become politically active – write letters to your representatives, participate in rallies, talk to your friends and neighbors.

14.4 A few final thoughts

In this book, I have tried to give a comprehensive overview of the climate change problem. Unlike what you might hear in the public debate, much of the science of climate change is extremely solid. There is no question that, when you add a greenhouse gas to the atmosphere, the planet will warm (Chapters 4 and 6). There is no question that human activities are increasing the amount of greenhouse gas in our atmosphere (Chapter 5). There is no question that the Earth is currently warming (Chapter 2), and it is warming about as much as you would expect from the addition of greenhouse gases (Chapter 7). This science is not new — much of it is a century or more old and it has stood the test of time. While it is possible for a scientific revolution to overturn mainstream climate science, that is highly unlikely.

Other aspects of the problem are, however, less certain. Quantitative projections of future climate change have large uncertainties because there are uncertainties in the physics as well as uncertainties in exactly how the world will economically evolve over the next century and beyond (Chapter 8). Moreover, this uncertainty is magnified by uncertainty in how this warming will impact humans and those aspects of the environment that we care about. However, one conclusion is clear: If climate change falls toward the upper end of the predicted range, we will truly be remaking

---

[6] This might make financial sense if you receive enough of a subsidy from the government.

the face of the planet, and the results will be dire, perhaps even catastrophic, for many of the world's inhabitants (Chapter 9).

We know how to solve this problem (Chapter 11, 12, and 14), but we don't know how hard and expensive it will be – and we won't know until we try. Paralyzed by this uncertainty, the world has made little progress in solving this problem, despite decades of warnings from scientists (Chapter 13).

I do not know what the future holds. But I do know that, if we're going to navigate the coupled problems of energy and climate, we're going to need people like you to get involved in all parts of the problem: the political, the economic, and the scientific. Given the enormous creativity and inventiveness of humans, there's no question that we *can* solve the problem. I encourage you to get involved to ensure that we do so.

Additional reading

N. Stern, *The Economics of Climate Change: The Stern Review* (Cambridge: Cambridge University Press), 2007; W. Nordhaus, *A Question of Balance: Weighing the Options on Global Warming Policies* (New Haven, CT: Yale University Press, 2008). These two economic analyses compare the costs and benefits of action on climate change. Both conclude that action is required, but they differ on how much action, mainly as a result of differences in the discount rate. Stern advocates strong action immediately, whereas Nordhaus advocates a slower ramping down on emissions.

P. Krugman, "Building a Green Economy," *New York Times Magazine*, April 7, 2010. This offers a clear and concise summary of the economics of climate change policy and a critique of cost–benefit calculations on climate change (download at http://www.nytimes.com/2010/04/11/magazine/11Economy-t.html).

A. E. Dessler and E. A. Parson, *The Science and Politics of Global Climate Change: A Guide to the Debate*, 2nd ed. (Cambridge: Cambridge University Press, 2010). Chapter 5 of that book covers the uncertainty question; it also outlines in detail the elements of an effective international response to climate change as well as how we might get there.

Problems

1. Explain the actions the world needs to take to limit global average warming to 2 °C. How do aerosols complicate our efforts to reduce radiative forcing?
2. A friend argues, "we must be certain climate change is a problem before we take action." Another friend argues, "we must take action if the slightest chance exists that climate change could be catastrophic." How do you determine which one is right? Which one (if either) do you judge to be correct?
3. How does irreversibility of the choices affect policy decisions?
4. a) How much would you pay each year to prevent the extinction of polar bears? How do you determine this value?

b) Imagine that your hometown is always at risk of being destroyed by some natural disaster (tornado, hurricane, earthquake). How much would you pay each year to eliminate the chance that the disaster would occur in that year? How do you determine this value?

5. a) Explain conceptually the role that discounting plays in determining the costs of climate change.

   b) If you change the discount rate from 0% to 4%, does this increase or decrease the present-value costs of climate impacts change?

6. In this chapter, we used a climate sensitivity of $0.75\,°C/(W/m^2)$ in order to calculate a radiative forcing limit for $2\,°C$ of $2.7\ W/m^2$. However, in Chapter 6 we noted that, considering the uncertainty, the sensitivity could lie in the $0.5–1.1\,°C/(W/m^2)$ range.

   a) What range of radiative forcing limits does this correspond to?

   b) If the climate sensitivity is $1.1\,°C/(W/m^2)$, what do you think our prospects are of meeting a $2\,°C$ limit?

7. Juries in criminal trials are given a standard of evidence that must be crossed in order to find a defendant guilty. What is it? What would the standard be if our society decided that the worse error is to acquit a guilty person?

8. a) Calculate the present value in 2010 of an impact with a cost of $10 per year every year between 2010 and 2110. Use a discount rate of 0% and 4%. Note that this problem is made much easier by using a spreadsheet program.

   b) Imagine that you could pay $500 today and avoid these impacts. Calculate a discount rate where you would be indifferent between paying $500 today or paying $10 per year.

References

Albritton, D. L., L. G. Meira Filho, U. Cubasch, X. Dai, Y. Ding, D. J. Griggs, B. Hewitson, J. T. Houghton, I. Isaksen, T. Karl, M. McFarland, V. P. Meleshko, J. F. B. Mitchell, M. Noguer, B. S. Nyenzi, M. Oppenheimer, J. E. Penner, S. Pollonais, T. Stocke, and K. E. Trenberth, "Technical Summary," in J. T. Houghton, Y. Ding, D. J. Griggs, M. Noguer, P. J. van der Linden, X. Dai, K. Maskell, and C. A. Johnson (eds.), *Climate Change 2001: The Scientific Basis*. Contribution of Working Group I to the Third Assessment Report of the Intergovernmental Panel on Climate Change (Cambridge: Cambridge University Press, 2001).

Bindoff, N. L., J. Willebrand, V. Artale, A, Cazenave, J. Gregory, S. Gulev, K. Hanawa, C. Le Quéré, S. Levitus, Y. Nojiri, C. K. Shum, L. D. Talley, and A. Unnikrishnan, "Observations: Oceanic Climate Change and Sea Level," in S. Solomon, D. Qin, M. Manning, Z. Chen, M. Marquis, K. B. Averyt, M. Tignor, and H. L. Miller (eds.), *Climate Change 2007: The Physical Science Basis*. Contribution of Working Group I to the Fourth Assessment Report of the Intergovernmental Panel on Climate Change (Cambridge: Cambridge University Press, 2007).

Deffeyes, K. S., *Beyond Oil: The View from Hubbert's Peak* (New York: Hill and Wang, 2006).

Denman, K. L., G. Brasseur, A. Chidthaisong, P. Ciais, P. M. Cox, R. E. Dickinson, D. Hauglustaine, C. Heinze, E. Holland, D. Jacob, U. Lohmann, S. Ramachandran, P. L. da Silva Dias, S. C. Wofsy, and X. Zhang, "Couplings between Changes in the Climate System and Biogeochemistry," in S. Solomon, D. Qin, M. Manning, Z. Chen, M. Marquis, K. B. Averyt, M. Tignor, and H. L. Miller (eds.), *Climate Change 2007: The Physical Science Basis*. Contribution of Working Group I to the Fourth Assessment Report of the Intergovernmental Panel on Climate Change (Cambridge: Cambridge University Press, 2007).

Dessler, A. E., and E. A. Parson, *The Science and Politics of Global Climate Change: A Guide to the Debate*, 2nd ed. (Cambridge: Cambridge University Press, 2010).

Forster, P., V. Ramaswamy, P. Artaxo, T. Berntsen, R. Betts, D. W. Fahey, J. Haywood, J. Lean, D. C. Lowe, G. Myhre, J. Nganga, R. Prinn, G. Raga, M. Schulz, and R. Van Dorland, "Changes in Atmospheric Constituents and in Radiative Forcing," in S. Solomon, D. Qin, M. Manning, Z. Chen, M. Marquis, K. B. Averyt, M. Tignor, and H. L. Miller (eds.), *Climate Change 2007: The Physical Science Basis*. Contribution of Working Group I to the Fourth Assessment Report of the Intergovernmental Panel on Climate Change (Cambridge: Cambridge University Press, 2007).

Houghton, J., "Madrid 1995: Diagnosing Climate Change," *Nature* 455 (2008): 737–738 (accessible online at http://www.nature.com/nature/journal/v455/n7214/full/455737a.html).

IPCC, "Summary for Policymakers," in S. Solomon, D. Qin, M. Manning, Z. Chen, M. Marquis, K. B. Averyt, M. Tignor, and H. L. Miller (eds.), *Climate Change 2007: The Physical Science Basis*. Contribution of Working Group I to the Fourth Assessment Report of the Intergovernmental Panel on Climate Change (Cambridge: Cambridge University Press, 2007).

IPCC, "Summary for Policymakers," in B. Metz, O. R. Davidson, P. R. Bosch, R. Dave, and L. A. Meyer (eds.), *Climate Change 2007: Mitigation*. Contribution of Working Group III to the Fourth Assessment Report of the Intergovernmental Panel on Climate Change (Cambridge: Cambridge University Press, 2007).

Jansen, E., J. Overpeck, K. R. Briffa, J.-C. Duplessy, F. Joos, V. Masson-Delmotte, D. Olago, B. Otto-Bliesner, W. R. Peltier, S. Rahmstorf, R. Ramesh, D. Raynaud, D. Rind, O. Solomina, R. Villalba, and D. Zhang, "Palaeoclimate," in S. Solomon, D. Qin, M. Manning, Z. Chen, M. Marquis, K. B. Averyt, M. Tignor, and H. L. Miller (eds.), *Climate Change 2007: The Physical Science Basis*. Contribution of Working Group I to the Fourth Assessment Report of the Intergovernmental Panel on Climate Change (Cambridge: Cambridge University Press, 2007).

Karl, T. R., S. J. Hassol, C. D. Miller, and W. L. Murray (eds.), *Temperature Trends in the Lower Atmosphere: Steps for Understanding and Reconciling Differences* (Washington, DC: U.S. Climate Change Science Program, 2006; available online at http://www.climatescience.gov/Library/sap/sap1-1/default.php).

Karl, T. R., J. M. Melillo, and T. C. Peterson (eds.), *Global Climate Change Impacts in the United States. Unified Synthesis Product* (Washington, DC: U. S. Climate Change Science Program, 2009; available online at http://www.climatescience .gov/Library/sap/usp/default.php).

Lemke, P., J. Ren, R. B. Alley, I. Allison, J. Carrasco, G. Flato, Y. Fujii, G. Kaser, P. Mote, R. H. Thomas, and T. Zhang, "Observations: Changes in Snow, Ice and Frozen Ground," in S. Solomon, D. Qin, M. Manning, Z. Chen, M. Marquis, K. B. Averyt, M. Tignor, and H. L. Miller (eds.), *Climate Change 2007: The Physical Science Basis*. Contribution of Working Group I to the Fourth Assessment Report of the Intergovernmental Panel on Climate Change Cambridge: Cambridge University Press, 2007).

Lisiecki, L. E., and M. E. Raymo, "A Pliocene–Pleistocene Stack of 57 Globally Distributed benthic delta O-18 Records," *Paleoceanography* 20 (2005): PA1003 (doi: 10.1029/2004PA001071).

Mehl, G. A., T. F. Stocker, W. D. Collins, P. Friedlingstein, A. T. Gaye, J. M. Gregory, A. Kitoh, R. Knutti, J. M. Murphy, A. Noda, S. C. B. Raper, I. G. Watterson, A. J. Weaver, and Z.-C. Zhao, "Global Climate Projections," in S. Solomon, D. Qin, M. Manning, Z. Chen, M. Marquis, K. B. Averyt, M. Tignor, and H. L. Miller (eds.), *Climate Change 2007: The Physical Science Basis*. Contribution of Working Group I to the Fourth Assessment Report of the Intergovernmental Panel on Climate Change (Cambridge: Cambridge University Press, 2007).

National Research Council, *Surface Temperature Reconstructions for the Last 2,000 Years*. Board on Atmospheric Sciences and Climate (Washington, DC: National Academies Press, 2006).

Nordhaus, W., *A Question of Balance: Weighing the Options on Global Warming Policies* (New Haven, CT: Yale University Press, 2008).

Petit, J. R., J. Jouzel, D. Raynaud, N. I. Barkov, J. M. Barnola, I. Basile, M. Bender, J. Chappellaz, M. Davis, G. Delaygue, M. Delmotte, V. M. Kotlyakov, M. Legrand, V. Y. Lipenkov, C. Lorius, L. Pepin, C. Ritz, E. Saltzman and M. Stievenard, "Climate and Atmospheric History of the Past 420,000 Years from the Vostok Ice Core, Antarctica," *Nature* 399 (1999): 429–436.

Royer, D. L., "CO_2-Forced Climate Thresholds during the Phanerozoic," *Geochimica et Cosmochimica Acta* 70 (2006): 5665–5675.

Solomon, S., D. Qin, M. Manning, R. B. Alley, T. Berntsen, N. L. Bindoff, Z. Chen, A. Chidthaisong, J. M. Gregory, G. C. Hegerl, M. Heimann, B. Hewitson, B. J. Hoskins, F. Joos, J. Jouzel, V. Kattsov, U. Lohmann, T. Matsuno, M. Molina, N. Nicholls, J. Overpeck, G. Raga, V. Ramaswamy, J. Ren, M. Rusticucci, R. Somerville, T. F. Stocker, P. Whetton, R. A. Wood, and D. Wratt, "Technical Summary," in S. Solomon, D. Qin, M. Manning, Z. Chen, M. Marquis, K. B. Averyt, M. Tignor, and H. L. Miller (eds.), *Climate Change 2007: The Physical Science Basis*. Contribution of Working Group I to the Fourth Assessment Report of the Intergovernmental Panel on Climate Change (Cambridge: Cambridge University Press, 2007).

Solomon, S., G.-K. Plattner, R. Knutti, and P. Friedlingstein, "Irreversible Climate Change due to Carbon Dioxide Emissions," *Proceedings of the National Academy of Sciences*, 106 (2009): 1704–1709.

Stanton, E. A., and F. Ackerman, *Florida and Climate Change: The Costs of Inaction* (Boston: Tufts University, 2007; available online at http://ase.tufts.edu/gdae/Pubs/rp/FloridaClimate.html).

Trenberth, K. E., P. D. Jones, P. Ambenje, R. Bojariu, D. Easterling, A. Klein Tank, D. Parker, F. Rahimzadeh, J. A. Renwick, M. Rusticucci, B. Soden, and P. Zhai, "Observations: Surface and Atmospheric Climate Change," in S. Solomon, D. Qin, M. Manning, Z. Chen, M. Marquis, K. B. Averyt, M. Tignor, and H. L. Miller (eds.), *Climate Change 2007: The Physical Science Basis*. Contribution of Working Group I to the Fourth Assessment Report of the Intergovernmental Panel on Climate Change (Cambridge: Cambridge University Press, 2007).

Union of Concerned Scientists, *Smoke, Mirrors, and Hot Air: How ExxonMobil Uses Big Tobacco's Tactics to Manufacture Uncertainty on Climate Science*. (Cambridge, MA: UCS, January 2007; available online at http://www.ucsusa.org/assets/documents/global_warming/exxon_report.pdf).

Velicogna, I., "Increasing Rates of Ice Mass Loss from the Greenland and Antarctic Ice Sheets Revealed by GRACE," *Geophysical Research Letters* 36 (2009): L19503 (doi:10.1029/2009GL040222).

Weart, S. R., *The Discovery of Global Warming*, 2nd ed. (Cambridge, MA: Harvard University Press, 2008; accessible online at http://www.aip.org/history/climate/index.htm).

Zachos, J., M. Pagani, L. Sloan, E. Thomas, and K. Billups, "Trends, Rhythms, and Aberrations in Global Climate 65 Ma to Present," *Science* 292 (2001): 686–693.

Index

2 °C limit, 170, 212, 222–223
    policy roadmap, 224–227
    radiative forcing limit, 223
    required emissions reductions, 223–224

A1, A2, A1FI, A1T. *See* emissions scenarios
abrupt climate change, 149
acidification of the ocean. *See* ocean acidification
acid rain, 203
    Reagan Administration response, 203
adaptation, 165–169
    connection to wealth, 167
    definition of success, 168
additionality, 193
aerosol indirect effect, 89
aerosols, 87, 223
    role in geoengineering, 179
affluence, 119, 121, 122, 224
    future trends, 123
    historical trends, 121
Agassiz, Louis, 198
air capture of carbon dioxide, 180
albedo, 50
argon, 62
Arrhenius, Svante, 198
atmospheric composition, 62

B1, B2. *See* emissions scenarios
biological carbon pump, 180
biomass energy, 173
blackbody radiation, 37
black carbon aerosols, 88, 223
Bush, George W., approach to climate policy,
    210–212

Callendar, Guy Stewart, 199
calorie, 34
cap and trade, 189–191
    carbon tax vs. cap and trade, 191–192
    escape valve, 192
carbohydrate, 65
carbon-12, 75
carbon-13, 75
carbon-14, 75, 76
carbonated water, 67
carbon capture and storage, 175
carbon cycle, 62, 79
    atmosphere-land biosphere exchange, 64
    atmosphere-ocean carbon exchange, 66

ocean acidification. *See* ocean acidification
atmosphere-rock carbon exchange, 69
    volcanoes. *See* volcanoes, emission of carbon
        dioxide
    human perturbations, 71
    role of continental drift, 104
carbon-cycle engineering, 180
carbon-cycle feedback, 97
carbon dioxide, 63
    abundance over the last 10,000 years, 72
    abundance over the last 250 years, 72
    abundance over the last half billion years, 70
    abundance since the late 1950s, 72
    from deforestation, 72
    dissolves into the ocean, 66
    from fossil fuels, 71
    future trends, 126
        beyond 2100, 130
    impact of long lifetime on climate change, 130
    isotopes, 75
    volcanic emissions, 69
carbonic acid, 66
carbon intensity, 120, 122, 123
    future trends, 123
    historical trends, 122
carbon sequestration. *See* carbon capture and
        storage
carbon tax, 186–189
    carbon tax vs. cap and trade, 191–192
CCS. *See* carbon capture and storage
Celsius temperature scale, 1
CFCs. *See* chlorofluorocarbons
chemical weathering, 69
chlorofluorocarbons, 63
    ozone depletion, 202
climate, 1
    definition, 1
    predictability, 131–133
climate change, 4
    abrupt climate changes. *See* abrupt climate
        change
    causes, 103
        changes in the solar constant, 104
        greenhouse gases, 109
        internal variability, 108
        movement of the continents, 103
        orbital variations, 105
    definition, 4
    importance of rate of change, 137

235